D1207187

FIXED POINTS AND TOPOLOGICAL DEGREE IN NONLINEAR ANALYSIS

MATHEMATICAL SURVEYS · *Number 11*

FIXED POINTS AND TOPOLOGICAL DEGREE IN NONLINEAR ANALYSIS

BY
JANE CRONIN

1964
AMERICAN MATHEMATICAL SOCIETY
190 Hope Street, Providence, Rhode Island

Library of Congress Catalog Number: 63-21550

PREFACE

Since this book is an introduction to the application of certain topological methods to nonlinear differential and integral equations, it is necessarily an incomplete account of each of these subjects. Only the topology which is needed will be introduced, and only those aspects of differential and integral equations which can be profitably studied by the use of the topological methods are discussed. The bibliography is intended to be representative, not complete.

I am greatly indebted to Professor Lamberto Cesari who read the entire manuscript and made a number of valuable suggestions and corrections concerning both the mathematics and the exposition in this book.

The preparation of this book was supported by the Air Force Office of Scientific Research. I wish to express my thanks to the Air Force Office of Scientific Research and also to the editorial staff of the American Mathematical Society for their work in preparing the manuscript for publication.

Jane Cronin.

FOREWORD

Our aim is to give a detailed description of fixed point theory and topological degree theory starting from elementary considerations and to explain how this theory is used in the study of nonlinear differential equations, ordinary and partial. The reader is expected to have a fair knowledge of advanced calculus, especially the point set theory of Euclidean n-space, but no further knowledge of topology is assumed. (Following this introduction is a list of definitions and notations from advanced calculus which will be used in the text.)

Why topological methods are used. It is easy to see that the solution $x(t)$ of the differential equation

$$\frac{d^2x}{dt^2} + x = 0,$$

such that $x(0) = 0$ and $dx(0)/dt = 1$, is $x = \sin t$. In finding this solution, we have obtained complete quantitative information, i.e., for any given value of t, we can find the corresponding value $x(t)$ by referring to a table of trigonometric functions. However, we have also obtained other useful information called qualitative or "in the large" information: we know that the values $x(t)$ are between -1 and $+1$ and that $x(t)$ has period 2π.

If all differential equations could be solved as easily, it would be unnecessary to introduce the distinction between quantitative and qualitative information concerning solutions. But this example is an exceptionally simple one. Most differential equations, especially nonlinear equations, must be studied with one technique to obtain quantitative information and by another to obtain qualitative information. If a method can be derived for finding the numerical values of the solution corresponding to given values of the independent variable, this method will usually not give us qualitative information. For example, to determine if the solution is periodic requires a different approach.

The topological methods we describe will give us qualitative information, e.g., information about the existence and stability of periodic solutions of ordinary differential equations and the existence of solutions of certain partial differential equations. However, they will gives no quantitative information, i.e., no information about how to compute any of the solution values. At best, we will get upper and lower bounds for solution values. The information that the topological methods give us is thus incomplete.

Nevertheless, topological methods are used because at present there are no other methods that yield as much qualitative information with so little effort. It is possible to envision a future in which topological methods will be supplanted by more efficient methods, but so far there is little encouragement for doing so. Such a venerable technique as the Poincaré-Bendixson Theorem has not been essentially improved in over sixty years although it is widely used. (The Poincaré-Bendixson Theorem is not one of the techniques which we will describe but as will be seen in Chapter II, it is intimately related to one of these techniques.)

The topological techniques to be developed. The two techniques to be developed, the fixed point theorem and local topological degree, are closely connected. The fixed point theorem has the advantage of being a comparatively elementary theorem (it can be proved without using any "topological machinery") which has many useful applications. The topological degree theory requires lengthier considerations for its development, but it has an important advantage over the fixed point theorem: it gives information about the number of distinct solutions, continuous families of solutions, and stability of solutions.

When to use topological techniques. We shall mostly be concerned with the question of *how* to apply topological techniques. The question of *when* to apply them is equally important. There can be no precise answer to the question, but we can formulate a rough rule. If we use an analytical method (like successive approximations), we establish existence and uniqueness of solution and a method (not necessarily practical) for computing the solution. If analytical means fail and especially if there seems to be no way to establish uniqueness, then we turn to the weaker question of establishing mere existence. For answering this weaker question, the topological methods (cruder and yielding less information than analytical methods) sometimes suffice. Thus topological methods should be regarded as a last resort or at least a later resort than analytical methods.

Summary of contents. In Chapter I, a definition of the local topological degree in Euclidean n-space is given, the basic properties of topological degree are derived, and some methods for computing the degree are described. Also the Brouwer Theorem (the fixed point theorem in Euclidean n-space) is obtained. In Chapter II, the techniques described in Chapter I are applied to some problems in ordinary differential equations: existence and stability of periodic and almost periodic solutions. In Chapter III, the Euclidean n-space techniques developed in Chapter I are extended to spaces of arbitrary dimension. We obtain the Leray-Schauder degree and the Schauder and Banach fixed point theorems for mappings in Banach space. We obtain also a combination of analytical and topological techniques which can be used to study local problems in Banach space. Finally in Chapter IV, the theory developed in Chapter III is applied to integral equations, partial differential equations and to some further problems on periodic solutions of ordinary differential equations.

The applications in Chapters II and IV are treated in varying detail. Existence of periodic solutions of nonautonomous ordinary differential equations is treated in complete detail. Stability of periodic solutions is treated in full detail except for the basic stability theorem of Lyapunov which is stated without proof. The other topics in Chapter II are similarly treated: certain theorems, particularly those from other disciplines, are stated without proof. In Chapter IV, an elaborate apparatus of theorems from analysis must be used in applying the Leray-Schauder theory and the Schauder Fixed Point Theorem. Some of these theorems have lengthy and complicated proofs and we restrict ourselves to giving references for these proofs.

The numbering of definitions, theorems, etc., is done independently in each chapter. Unless otherwise stated, references to a numbered item means that item in the chapter in which the reference is made. E.g., if in Chapter II, reference is made to Theorem (3.8), that means Theorem (3.8) in Chapter II.

Some Terminology and Notation Used in This Text

▌ denotes end of proof.

nasc is abbreviation for necessary and sufficient condition.

Set notation

$\in$: element of
$\notin$: not an element of
$\bigcup$: union
$\bigcap$: intersection
$-$: difference
$\varnothing$: null set
$\subset$: contained in
A^c: complement of set A
$A \times B$: Cartesian product of sets A and B, i.e., the set of all ordered pairs (a, b) where $a \in A$, $b \in B$.

The set of elements having property P is denoted by:

$$[x \,/\, x \text{ has property } P].$$

If a, b are real numbers such that $a < b$, the following notation is used to indicate the various intervals:

$$[a, b] = [x \text{ real} \,/\, a \leqq x \leqq b],$$
$$[a, b) = [x \text{ real} \,/\, a \leqq x < b],$$
$$(a, b] = [x \text{ real} \,/\, a < x \leqq b],$$
$$(a, b) = [x \text{ real} \,/\, a < x < b].$$

$[a, b]$ is called a closed interval and (a, b) is called an open interval. R^n denotes real Euclidean n-space, i.e., the collection of n-tuples of real numbers

$(x_1, \cdots, x_n)$. The elements of R^n will be denoted by single letters, $p, q, \cdots$ when this is possible. If $p = (p_1, \cdots, p_n)$ and λ is a real number, then

$$\lambda p = (\lambda p_1, \cdots, \lambda p_n).$$

In particular,

$$(-1)p = (-p_1, \cdots, -p_n).$$

If $p = (p_1, \cdots, p_n)$ and $q = (q_1, \cdots, q_n)$, then

$$p + q = (p_1 + q_1, \cdots, p_n + q_n).$$

The *distance between points p and q*, denoted by $|p - q|$, is:

$$|p - q| = [(p_1 - q_1)^2 + \cdots + (p_n - q_n)^2]^{1/2}.$$

If $p \in R^n$, the neighborhoods of p are the sets:

$$N_\varepsilon(p) = [q \,/\, |p - q| < \varepsilon]$$

where ε is an arbitrary positive number. A set O in R^n is *open* if for each point $p \in O$, there is a neighborhood $N_\varepsilon(p)$ of p such that $N_\varepsilon(p) \subset O$. A set F in R^n is *closed* if F^c is open. A point p is a *limit point* (*cluster point, accumulation point*) of a set E in R^n if each neighborhood of p contains at least one point of E distinct from p. (Point p may or may not be in E.) If $D \subset R^n$, a *boundary point* of D is a point p such that each neighborhood of p contains a point of D and a point of D^c. (A boundary point of D may or may not be in D.) The collection of boundary points of D is denoted by D'. The set $D \cup D'$ (also denoted by $\bar{D}$) is called the *closure* of D. The set $D \cup D'$ is a closed set. If there is a neighborhood $N_\varepsilon(p)$ of a point p such that $N_\varepsilon(p)$ is contained in a set E, then p is an *interior point* of E.

A *metric space* is a collection M of points $p, q, \cdots$ for which a function ρ from $M \times M$ into the non-negative real numbers is defined such that:

(1) $\rho(p, q) > 0$ if and only if $p \neq q$;
(2) $\rho(p, q) = \rho(q, p)$;
(3) $\rho(p, r) \leqq \rho(p, q) + \rho(q, r)$.

All the concepts defined for R^n, i.e., neighborhood, open set, closed set, etc., may be defined for a metric space by using the same definitions given for R^n only with $|p - q|$ replaced by $\rho(p, q)$.

A *separable metric* space is a metric space M such that there is a denumerable subset $[x_n]$ of M such that the closure of $[x_n]$ is M.

A *connected* set in a metric space M is a set S such that S is not the union of two disjoint nonempty sets A and B which are contained in disjoint open sets.

A *component* of an open set O in a metric space is a maximal connected subset of O.

A *locally connected* metric space is a metric space M such that if $p \in M$ and $N_\varepsilon(p)$ is a neighborhood of p then there is a connected neighborhood of p which is contained in $N_\varepsilon(p)$.

If A is a subset of a metric space M, the *diameter* of A is $\text{lub}_{a,b \in A}\, \rho(a, b)$.

If A, B are disjoint subsets of a metric space M, then the *distance between A and B* is

$$\underset{a \in A;\, b \in B}{\text{glb}} \quad \underset{(a,\, b) \in A \times B}{\rho(a, b)}$$

and is denoted by $d(A, B)$.

A function f from a metric space M_1 into a metric space M_2 is 1-1 if p, $q \in M_1$ and $p \neq q$ imply $f(p) \neq f(q)$.

Function f from metric space M_1 into metric space M_2 is *continuous* if for each open set O in M_2, the set $f^{-1}(O)$, where f^{-1} is the inverse of f, is also open.

If f is a 1-1 continuous function from M_1 into M_2 and if f^{-1} is continuous, then f is a *homeomorphism* from M_1 onto $f(M_1)$.

If f is a function from metric space M_1 into metric space M_2 and if A is a subset of M_1, then f/A denotes the function f regarded only on A, i.e., the function with domain A such that if $x \in A$, then the functional value is $f(x)$.

TABLE OF CONTENTS

Chapter I. TOPOLOGICAL TECHNIQUES IN EUCLIDEAN n-SPACE

0. Introduction 1
1. The fixed point theorem 1
2. The order of a point relative to a cycle: cells, chains, and cycles; orientation of R^n; intersection numbers; order of a point relative to a cycle 2
3. The order of a point relative to a continuous image of z^{n-1} . 16
4. Properties of $v[\phi, \bar{K}, p]$ 25
5. The local degree relative to a complex 26
6. The local degree relative to the closure of a bounded open set . 30
7. The local degree as a lower bound for the number of solutions . 32
8. A product theorem for local degree 36
9. Computation of the local degree 37
10. A reduction theorem and an in-the-large implicit function theorem 50
11. A proof of the fixed point theorem 52
12. The index of a fixed point 52
13. The index of a vector field 53
14. Generalizations 54

Chapter II. APPLICATIONS TO ORDINARY DIFFERENTIAL EQUATIONS

1. Some existence theorems for differential equations . . 56
2. Linear systems 63
3. Existence of periodic solutions of nonautonomous quasilinear systems 64
4. Some stability theory 70
5. Stability of periodic solutions of nonautonomous quasilinear systems 75
6. Some examples of nonautonomous quasilinear systems . . 85
7. Almost periodic solutions of nonautonomous quasilinear systems 96
8. Periodic solutions of autonomous quasilinear systems . . 105
9. Periodic solutions of systems with a "large" nonlinearity . 109

Chapter III. TOPOLOGICAL TECHNIQUES IN FUNCTION SPACE

1. Introduction 119
2. Some linear space theory 120
3. Examples which show that a fixed point theorem and a definition of local degree cannot be obtained for arbitrary continuous transformations from a Banach space into a Banach space: Kakutani's example; Leray's example 124

4. Compact transformations: properties of compact transformations; Schauder Fixed Point Theorem; Schaefer's Theorem . . 130
5. Definition and properties of the Leray-Schauder degree . . 134
6. Proof of the Schauder Theorem using the Leray-Schauder degree 139
7. Computation of the Leray-Schauder degree 139
8. A partially analytic approach: contraction mappings; Banach Fixed Point Theorem; some further Banach space theory; local study 140

Chapter IV. APPLICATIONS TO INTEGRAL EQUATIONS, PARTIAL DIFFERENTIAL EQUATIONS AND ORDINARY DIFFERENTIAL EQUATIONS WITH LARGE NONLINEARITIES

1. Introduction 151
2. Integral equations 152
3. Problems in partial differential equations 156

Elliptic differential equations

4. Statement of the Leray-Schauder-Nirenberg result 156
5. The Banach spaces in which the Leray-Schauder-Nirenberg result is formulated 157
6. The Schauder Existence Theorem 161
7. The Leray-Schauder method 162
8. The Nirenberg method 167
9. Some other work on elliptic equations 170
10. Local study of elliptic differential equations 171

Parabolic differential equations

11. An analog of the Schauder Existence Theorem 176
12. Some results for quasilinear parabolic equations 178

Hyperbolic differential equations

13. The Cauchy problem 180
14. Mixed problems 180

Ordinary differential equations

15. The Cesari method for ordinary nonlinear equations . . . 180

BIBLIOGRAPHY 186
INDEX 194

CHAPTER I

Topological Techniques in Euclidean n-space

Introduction. The two topological concepts we use are the fixed point theorem and local topological degree (hereafter to be termed local degree or degree). The fixed point theorem is couched in simple terms and we will be able to state it with practically no introduction. But describing the local degree theory is lengthier. The description to be given (which is essentially the singular homology definition in Alexandroff-Hopf [**1**]) was chosen because it includes the "order" viewpoint and the "covering number" viewpoint and therefore seems to be the most suggestive of ways to compute the local degree. (For use in applications, we will have to make such computations.)

Nevertheless we describe briefly other methods of defining local degree. The earliest version is the definition in terms of the Kronecker integral, i.e., the degree is defined to be equal to a certain integral and the standard properties of the local degree are then proved. (See Alexandroff-Hopf [**1**, pp. 465–467] for discussion and references.) This definition holds only for differentiable mappings and does not give much basis for computing the degree. (Computing the integral itself is generally difficult.) Also we cannot obtain from this definition a theorem relating the degree and the number of solutions of a corresponding equation. Such a theorem is of considerable importance in applications. Another definition based entirely upon real analysis is given by Nagumo [**1**]. This definition is not long but would require some extension if it were to make a satisfactory basis for developing methods of computing the degree. A definition of degree in terms of cohomology is given by Rado and Reichelderfer [**1**]. From the point of view of a topologist, this is a more desirable definition than the one we give. It has, however, the disadvantages that it does not have as clear a geometric meaning as the Alexandroff-Hopf definition that we use (see Rado and Reichelderfer [**1**, p. 120, footnote 1]) and also it yields fewer suggestions for computing the degree. If the degree is defined only for mappings from the plane into itself, a much shorter definition can be given (see Alexandroff-Hopf [**1**, p. 464]). But it is important for later applications that our definition be applicable to mappings in Euclidean space of arbitrary finite dimension.

1. **The fixed point theorem.** The fixed point theorem says that if f is a continuous mapping of a solid sphere into itself, then f takes at least one point into itself, i.e., f leaves at least one point fixed. In the precise statement of the theorem which follows, we include a slightly wider class of sets than spheres.

DEFINITION. A *topological mapping* g of a set $E \subset R^n$ into R^n is a 1-1 continuous mapping such that g^{-1} is also continuous.

NOTATION. Let σ^n denote the solid unit sphere in R^n, i.e.,

$$\sigma^n = [(x_1, \cdots, x_n) \mid x_1^2 + \cdots + x_n^2 \leqq 1].$$

(1.1) BROUWER FIXED POINT THEOREM. *Let $B^n = g(\sigma^n)$ where g is a topological mapping. Let f be a continuous mapping of B^n into itself. Then there is an element x of B^n such that $x = f(x)$, i.e., the mapping f has a fixed point.*

This theorem is intuitively reasonable in the sense that any mapping f that one considers clearly does have such a fixed point. However, this observation is far from a proof of the theorem.

We will postpone proving the theorem until we have defined the local degree. The comparatively sophisticated degree theory will make possible a very short proof of the fixed point theorem. There are elementary proofs of the theorem, i.e., proofs which require few facts about topology (see Alexandroff-Hopf [**1**, p. 376 ff.]). As might be expected, an elementary proof is somewhat longer.

The fixed point theorem illustrates some typical traits of qualitative techniques. In terms of analysis, the theorem tells us that under certain circumstances the equation in R^n,

$$x - f(x) = 0, \tag{1.2}$$

has at least one solution. The theorem has two disadvantages: first it gives no information about how to find (i.e., compute) a solution of (1.2); secondly, it gives no information about how many solutions (1.2) has beyond the statement that (1.2) has at least one solution. Equation (1.2) may have just one solution or it may have an infinite set of solutions. For example: if $B^n = \sigma^n$ and f is the identity mapping, then every point $x \in \sigma^n$ is a solution of $x - f(x) = 0$; if $B^n = \sigma^2$ and f is a rotation of π radians, the only solution of (1.2) is (0, 0).

The first disadvantage, that no method for computing the solution is given, is inherent in the qualitative approach. A qualitative method usually establishes only the existence of a solution. When a qualitative method is used, we must expect to regard the computation of the solution as a separate problem. The second disadvantage, that no indication of the number of solutions is given, will be remedied when we have developed the local degree. At the cost of developing some topological "machinery," we will obtain estimates on the number of solutions.

2. **The order of a point relative to a cycle.** From the viewpoint of the analyst who wishes to apply degree theory, the local degree is a kind of estimate of the number of points mapped into a given point by a given mapping. If f is a continuous mapping from Euclidean n-space R^n into R^n, then the degree of f at point $p \in R^n$ is to be an estimate of the number of

points mapped by f into point p (these points are called p-points). We will require that this estimate remain unchanged or invariant if the mapping f or the point p is varied slightly. (As will be seen later, this condition is of crucial importance in applications of degree in analysis.) This important requirement of invariance unfortunately excludes the possibility of making the definition of degree the simplest possible one, i.e., defining the degree of f at point p as equal to the number of p-points of f. For suppose f is the mapping of R^1 into R^1 defined by

$$f: x \to x'$$

where $x' = x^2$. Then, if the degree were simply the number of p-points, the degree of f at 0 would be one. However, if our mapping f were changed to

$$f_\varepsilon : x \to x''$$

where $x'' = x^2 + \varepsilon$ and ε is a small positive number, then the degree of f at 0 would be zero no matter how small ε were chosen. If ε were a negative number, no matter how close to zero, the degree of f_ε at 0 would be two. Thus our requirement of invariance of the degree under small changes of f could not be satisfied.

To remedy this, we count the points mapped by f into p in a special way. Each p-point is counted with a plus or minus sign depending on whether the mapping f maps the points near the p-point so that directions are preserved or reversed. For example, if f is the mapping from R^1 into R^1:

$$f: x \to x^2 - \varepsilon$$

where ε is a positive number, the mapping f takes the interval $[0, \delta]$ where $\delta^2 > \varepsilon$ onto the interval $[-\varepsilon, \delta^2 - \varepsilon]$ without changing directions on $[0, \delta]$. On the other hand, the interval $[-\delta, 0]$ which is also mapped onto $[-\varepsilon, \delta^2 - \varepsilon]$ is "flipped over" in the process of being mapped. Roughly speaking, its direction is reversed. The 0-point $\sqrt{\varepsilon}$ is counted with a plus sign and the 0-point $-\sqrt{\varepsilon}$ is counted with a negative sign; hence we say that the degree of f at 0 is zero. With this definition, the degree is a crude estimate of the number of p-points—crude in that if the degree is nonzero, then there exists at least one p-point but if the degree is zero then there may be p-points (as in the example described) or there may be none at all.

The preceding is clearly far from a precise definition. We have not even said exactly what is meant by "flipping over" or reversing directions, much less given any indication of how this is done in an n-space R^n where $n > 1$.

Our first purpose is to show that the rough description of degree given above can be made into a precise definition for continuous mappings in Euclidean n-spaces. Then we define exactly what is to be meant by changing the mapping continuously, and prove that the degree is invariant under such changes. This fairly lengthy procedure will occupy the next five sections of Chapter I.

In order to define the degree of a mapping at a point, we will need some "combinatorial" concepts. The purpose of introducing these concepts is to make possible the definition of the order of a point relative to a cycle. This notion of order can be regarded as the simplest version of the local degree.

CELLS, CHAINS, AND CYCLES.

DEFINITION. A *convex* set $E \subset R^n$ is a set with the property: if $a, b \in E$ then all points $\lambda a + \mu b$ where $0 \leqq \lambda \leqq 1$, $0 \leqq \mu \leqq 1$, and $\lambda + \mu = 1$, are contained in E.

REMARK. It follows from the definition that if $\{C_v\}$ is a collection (finite or infinite) of convex sets, then $\bigcap_v C_v$ is convex.

DEFINITION. Let $a_0, a_1, \cdots, a_q$ be a finite set of distinct points in R^n. The *convex hull* of $a_0, a_1, \cdots, a_q$ is the convex set which is the intersection of all the convex sets which contain $a_0, a_1, \cdots, a_q$. We denote the convex hull by $\overline{a_0 a_1 \cdots a_q}$.

DEFINITION. Let U be an open set in R^n. (In particular U may be R^n itself.) Let $a_0, a_1, \cdots, a_q$ be a set of $(q + 1)$ distinct points in U such that $\overline{a_0 a_1 \cdots a_q} \subset U$. The set $\overline{a_0 a_1 \cdots a_q}$ is a *q-cell in U* or a *cell of order q in U*. The q-cell will sometimes be denoted by $\bar{x}^q$. (Also, if the superscript q is not needed, it will be omitted.) The points $a_0, a_1, \cdots, a_q$ are the *vertices* of $\bar{x}^q$.

DEFINITION. If $a_{k_1}, \cdots, a_{k_m}$ is a subset of $a_0, a_1, \cdots, a_q$ such that $\overline{a_{k_1} \cdots a_{k_m}}$ is a subset of the boundary of $\overline{a_0 a_1 \cdots a_q}$ then $\overline{a_{k_1} \cdots a_{k_m}}$ is a *side* of $\overline{a_0 a_1 \cdots a_q}$.

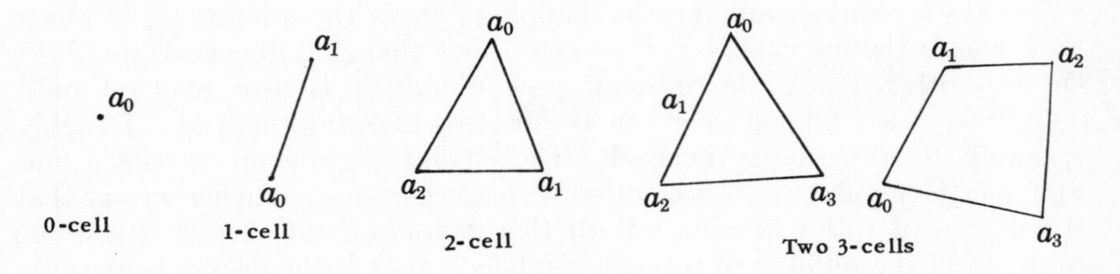

FIGURE 1. SOME CELLS IN R^2

DEFINITION. Consider the collection of all orderings of the vertices of a cell $\bar{x}^q$. Two orderings are *equivalent* if one can be obtained from the other by an even permutation of the vertices.

It is easy to show that this is a genuine equivalence relation, i.e., the relation is symmetric, reflexive, and transitive. Also just two equivalence classes of orderings are obtained.

DEFINITION. The two equivalence classes of orderings are called the *orientations* of $\bar{x}^q$. An *oriented cell in an open set U* is a cell $\bar{x}^q$ in U with one of the orientations specified. (More precisely, an oriented cell is a set with two elements: the cell $\bar{x}^q$ and one of the orientations.) Corresponding to each cell $\bar{x}^q$, there are two oriented cells which are denoted by $+x^q$ (or just x^q) and $-x^q$. These are called the *positively oriented cell* and the *negatively oriented cell* or more briefly the positive cell and the negative cell. If $a_0, a_1, \cdots, a_q$ are the vertices of $\bar{x}^q$, and if $a_0 a_1 \cdots a_q$ is an ordering in the positive orientation, the oriented cell x^q is sometimes denoted by $(a_0 a_1 \cdots a_q)$.

Note that the terms "positively oriented" and "negatively oriented" are assigned in an entirely arbitrary way. There is no real reason for calling a particular orientation the positive one.

If we try to apply the above definition of orientation to a 0-cell (which consists of just one point), then only one orientation is obtained for the cell. Consequently it would seem purposeless to introduce an orientation for a 0-cell. However, if a definition of orientation is not made for the 0-cell, then in any reference to orientation, we would have to treat the 0-cell as a special case. Consequently we make the following definition.

DEFINITION. An *oriented* 0-*cell* is a 0-cell with the only *possible orientation* specified. This orientation is called the positive orientation.

The geometric notion of a cell and the algebraic notion of "counting" the cell a certain number of times are combined in the concept of chain in which we associate with each one of a finite set of cells a "coefficient," i.e., a positive or negative integer, and in this association it is required that "multiplying" a cell by a negative integer $-n$ is "the same" as changing the orientation of the cell and "multiplying" the cell with reversed orientation by $+n$. We make this idea precise with the following definition.

DEFINITION. Let C^q be the collection of oriented q-cells in U where q is fixed. A *q-chain on U* is a function c^q with domain C^q and range a subset of the integers and with the properties:

(1) if $q > 0$,

$$c^q(-x^q) = -c^q(x^q)$$

for all $x^q \in C^q$;

(2) $c^q(x^q) \neq 0$ for at most a finite number of elements x^q of C^q.

(If it is not needed, the superscript q in c^q will be omitted. Also the phrase "on U" will often be omitted.)

NOTATION. The chain which has the value $+n$ on an oriented cell x^q, the value $-n$ on $-x^q$, and is zero elsewhere will be denoted by nx^q or if $n = 1$, the chain may be simply denoted by x^q. In general, the q-chain c will be denoted by $\sum_{i=1}^{m} t^i x_i$ where x_i is the oriented q-cell such that $c(x_i) = t^i$ and $x_1, \cdots, x_m, -x_1, \cdots, -x_m$ are the oriented q-cells for which the functional value is nonzero.

DEFINITION. If c_1, c_2 are the q-chains $\sum_i t^i x_i$, $\sum_j u^j x_j$, respectively, the *sum of* c_1 *and* c_2, denoted by $c_1 + c_2$, is the q-chain $\sum_k (t^k + u^k)x_k$ where the x_k are those oriented q-cells such that $t^k + u^k = c_1(x_k) + c_2(x_k) \neq 0$. If a is an integer and c is the q-chain $\sum t^i x_i$, then ac is defined to be the q-chain $\sum at^i x_i$. In particular if $a = -1$, then ac is denoted by $-c$. If c is the q-chain for which all the functional values are zero, then c is called the null q-chain and is denoted by θ_q.

DEFINITION. If $x^q = (a_0 a_1 \cdots a_q)$ is the q-chain which has the value 1 on x^q and -1 on $-x^q$ and is zero elsewhere, the *boundary of* x^q, denoted by $b(x^q)$, is the $(q - 1)$-chain

$$b(x_q) = \sum_{i=0}^{q} (-1)^i x_i^{q-1}$$

where $x_i^{q-1} = (a_0 a_1 \cdots \hat{a}_i \cdots a_q)$, i.e., the vertices of $\bar{x}^q$ with the same ordering as in x^q and with the vertex a_i omitted.

To make this definition valid, we must show that it is independent of the particular ordering $a_0 a_1 \cdots a_q$ in the orientation. Suppose $a_0 a_1 \cdots a_q$ and $a_{i_0} a_{i_1} \cdots a_{i_q}$ are orderings in the same orientation so that

$$(a_{i_0} a_{i_1} \cdots a_{i_q}) = (a_0 a_1 \cdots a_q).$$

If $a_{i_j} = a_k$, it is sufficient to show that

$$(-1)^k (a_0 a_1 \cdots \hat{a}_k \cdots a_q) = (-1)^{i_j} (a_{i_0} a_{i_1} \cdots \hat{a}_{i_j} \cdots a_{q_j}).$$

But

$$(a_0 a_1 \cdots a_k \cdots a_q) = (-1)^k (a_k a_0 a_1 \cdots \hat{a}_k \cdots a_q)$$

and

$$(a_{i_0} a_{i_1} \cdots a_{i_j} \cdots a_q) = (-1)^{i_j} (a_{i_j} a_{i_0} a_{i_1} \cdots \hat{a}_{i_j} \cdots a_q)$$

or

$$(-1)^k (a_k a_0 a_1 \cdots \hat{a}_k \cdots a_q) = (-1)^{i_j} (a_{i_j} a_{i_0} a_{i_1} \cdots \hat{a}_{i_j} \cdots a_q).$$

Since $a_k = a_{i_j}$, we have:

$$(-1)^k (a_0 a_1 \cdots \hat{a}_k \cdots a_q) = (-1)^{i_j} (a_{i_0} a_{i_1} \cdots \hat{a}_{i_j} \cdots a_q).$$

For examples of the boundary of a chain, see Figure 2.

REMARK. A definition of boundary for a 0-chain can be introduced but since we do not need it, we omit it.

DEFINITION. If c is the q-chain $\sum_i t^i x_i$ where $q \geqq 1$, the boundary of c, denoted by $b(c)$, is:

$$b(c) = \sum_i t^i b(x_i).$$

REMARK. From these definitions, it follows that:

$$b(c_1 + c_2) = b(c_1) + b(c_2)$$
$$b(-c) = -b(c)$$
$$b(\theta_q) = \theta_{q-1}.$$

DEFINITION. The q-chain c is a *q-cycle* if $b(c) = \theta_{q-1}$.

NOTATION. A q-cycle will be denoted by z^q.

THEOREM. *If $q \geqq 2$, then $b(x^q)$ is a $(q-1)$-cycle.*

PROOF. Let

$$x^q = (a_0 a_1 \cdots a_q),$$
$$x_i^{q-1} = (a_0 a_1 \cdots \hat{a}_i \cdots a_q),$$
$$x_j^{q-1} = (a_0 a_1 \cdots \hat{a}_j \cdots a_q),$$
$$x_{ij}^{q-2} = (a_0 a_1 \cdots \hat{a}_i \cdots \hat{a}_j \cdots a_q).$$

It is sufficient to show that the value of the chain $b[b(x^q)]$ on x_{ij}^{q-2} is zero. First the chain $b(x^q)$ has the value $(-1)^m$ on x_m^{q-1} for $m = i, j$. If $i < j$, then $b(x_i^{q-1})$ has value $(-1)^{j-1}$ on x_{ij}^{q-2} and $b(x_j^{q-1})$ has value $(-1)^i$ on x_{ij}^{q-2}. Thus if $i < j$, the value of $b[b(x^q)]$ on x_{ij}^{q-2} is:

$$[(-1)^i(-1)^{j-1} + (-1)^j(-1)^i] = 0. \blacksquare$$

DEFINITION. Let U, V be open sets in R^n. The q-cycle z^q on U is a *bounding cycle in V* if there is a $(q+1)$-chain c^{q+1} on V such that:

$$b(c^{q+1}) = z^q.$$

(2.1) THEOREM. *If z^q is a cycle on U, an open set in R^n, and if $q \geqq 1$, then z^q is a bounding cycle in R^n.*

PROOF. Let $z^q = \sum_{i=1}^m t^i x_i^q$. Let a be a point in R^n which is not a vertex of any of the cells $x_1^q, \cdots, x_m^q$. If $x_i^q = (a_0 a_1 \cdots a_q)$, let $(a x_i^q)$ denote the oriented cell $(a a_0 a_1 \cdots a_q)$ and let $a z^q$ denote the chain

$$\sum_{i=1}^m t^i (a x_i^q).$$

In particular, if $z^q = \theta_q$, then $a z^q$ denotes θ_{q+1}.

If $b(x_i^q)$ is $\sum_{j=0}^n (-1)^j x_j^{q-1}$, let $ab(x_i^q)$ denote the chain $\sum_{j=0}^n (-1)^j (a x_j^{q-1})$. Then

$$\begin{aligned} b[(az^q)] &= \sum_{i=1}^m t^i b(a x_i^q) \\ &= \sum_{i=1}^m t^i [x_i^q - ab(x_i^q)] \\ &= \sum_{i=1}^m t^i x_i^q - ab(z^q) \\ &= \sum_{i=1}^m t^i x_i^q = z^q. \blacksquare \end{aligned}$$

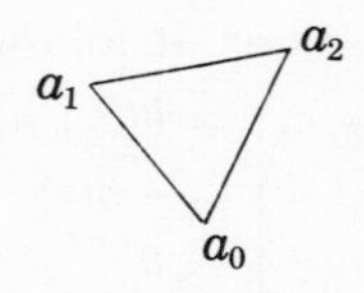

(i) $b(a_0a_1a_2) = (a_0a_1) - (a_0a_2) + (a_1a_2)$.

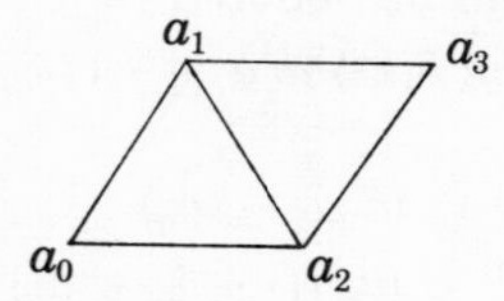

(ii) $$b[(a_0a_1a_2) + (a_1a_3a_2)] = (a_0a_1) - (a_0a_2) + (a_1a_2) - (a_1a_2) + (a_1a_3) + (a_3a_2)$$
$$= (a_0a_1) - (a_0a_2) + (a_1a_3) + (a_3a_2).$$

Note that (a_1a_2) appears in the boundary of each of the 2-cells and "cancels out."

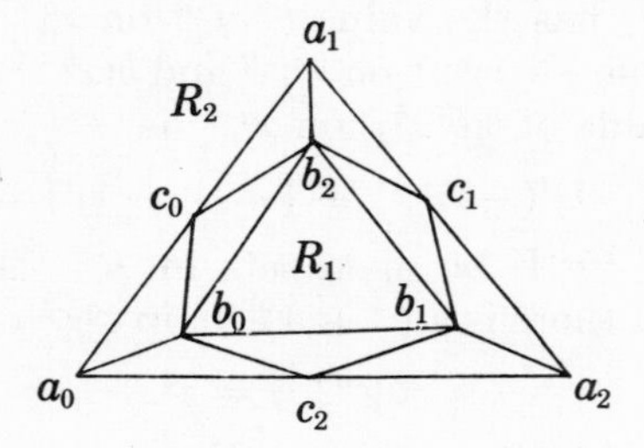

(iii) $$b[(a_0c_2b_0) + (b_0c_2b_1) + (b_1c_1a_2) + (b_1a_2c_1) + (b_1c_1b_2) + (b_2c_1a_1) + (b_2a_1c_0) + (b_2c_0b_0) + (b_0c_0a_0)]$$
$$= (a_0c_2) + (c_2a_2) + (a_2c_1) + (c_1a_1) + (a_1c_0) + (c_0a_0) - (b_0b_1) - (b_1b_2) - (b_2b_0).$$

Thus if U is the open set bounded by circles R_1 and R_2, then

$$(a_0c_2) + (c_2a_2) + (a_2c_1) + (c_1a_1) + (a_1c_0) + (c_0a_0) \sim [(b_0b_1) + (b_1b_2) + (b_2b_0)]$$

in U. But neither of these 1-chains is homologous to θ_1 in U.

FIGURE 2. SOME BOUNDARIES

DEFINITION. Suppose z_1^q, z_2^q are cycles in U. Then z_1^q is *homologous in U* to z_2^q if there exists a chain c^{q+1} in U such that

$$b(c^{q+1}) = z_1^q - z_2^q.$$

In particular if there exists c^{q+1} such that

$$b(c^{q+1}) = z^q,$$

then z^q is homologous to θ_q. We write $z_1^q \sim z_2^q$ in U to indicate that z_1^q is homologous in U to z_2^q.

REMARK. From the definitions we have made, it follows that the set of q-chains is an abelian group and that the boundary operator b is a homomorphism from the group of q-chains into the group of $(q - 1)$-chains. These facts are of fundamental importance in combinatorial topology.

ORIENTATION OF R^n. Our next task is to give a formal definition of orientation of R^n. This includes and makes precise the familiar intuitive notions of clockwise and counterclockwise curves in the plane and left-hand and right-hand coordinate systems in 3-space.

DEFINITION. The points $p_0, p_1, \cdots, p_q$ in R^n are *linearly independent* if they are not contained in a Euclidean $(q - 1)$-space R^{q-1}. (An equivalent definition is: the vectors $\overline{p_0p_1}, \overline{p_0p_2}, \cdots, \overline{p_0p_q}$ are linearly independent.)

DEFINITION. A cell $\bar{x}^q$ is a *q-simplex* if the $(q + 1)$ vertices of $\bar{x}^q$ are linearly independent. (In Figure 1, the 0-cell, the 1-cell, and the 2-cell are a 0-simplex, a 1-simplex, and a 2-simplex, respectively. Neither of the 3-cells is a 3-simplex.)

DEFINITION. An *affine mapping*

$$M:(x_1, \cdots, x_n) \to (x'_1, \cdots, x'_n)$$

of R^n into R^n is a mapping defined by the equations:

$$x'_i = \sum_{j=1}^{n} a_{ij}x_j + b_i \qquad (i = 1, \cdots, n)$$

where a_{ij} and b_i are constants $(j = 1, \cdots, n)$. The *determinant of M* is the determinant $|a_{ij}|$.

We shall deal only with nonsingular affine mappings, i.e., affine mappings with nonzero determinants. A nonsingular affine mapping is 1-1 and maps R^n onto R^n. Throughout the following discussion, "affine mapping" means nonsingular affine mapping.

DEFINITION. An affine mapping of R^n onto itself is *positive* [*negative*] if its determinant is positive [negative].

LEMMA. *Let $y_1^n = (a_0a_1 \cdots a_n)$ and y_2^n be oriented n-simplexes in R^n and let L be an affine mapping of R^n onto itself such that the points $L(a_0)$, $L(a_1)$, $\cdots$, $L(a_n)$ are the vertices of $\bar{y}_2^n$. Suppose that for some ordering $a_{r_0}, a_{r_1}, \cdots$, a_{r_n} of the vertices $a_0, a_1, \cdots, a_n$ which is in the positive orientation (of $\bar{y}_1^n$) the ordering $L(a_{r_0}), L(a_{r_1}), \cdots, L(a_{r_n})$ is in the positive orientation (of $\bar{y}_2^n$). Then if $a_{s_0}, a_{s_1}, \cdots, a_{s_n}$ is any ordering in the positive orientation of $\bar{y}_1^n$, the ordering $L(a_{s_0}), L(a_{s_1}), \cdots, L(a_{s_n})$ is in the positive orientation of $\bar{y}_2^n$.*

PROOF. Follows from the definition of orientation. The same permutation which takes $a_{r_0}, a_{r_1}, \cdots, a_{r_n}$ into $a_{s_0}, a_{s_1}, \cdots, a_{s_n}$ takes $L(a_{r_0}), \cdots, L(a_{r_n})$ into $L(a_{s_0}), L(a_{s_1}), \cdots, L(a_{s_n})$. ▌

This lemma permits us to formulate the definition:

Definition. If $y_1^n = (a_0 a_1 \cdots a_n)$ is an oriented simplex in R^n and if L is an affine mapping of R^n onto itself such that $(L(a_0) \cdots L(a_n))$ is y_2^n, an oriented simplex in R^n, then L *maps oriented simplex* y_1^n *onto oriented simplex* y_2^n.

Lemma. *Let* $\{L_\nu\}$ *be the collection of affine mappings which map* R^n *onto itself and which map an oriented simplex* $y_1^n = (a_0 a_1 \cdots a_n)$ *onto an oriented simplex* y_2^n. *Then the mappings* L_ν *are all positive or all negative.*

Proof. Suppose the statement is not true. Then there exist affine mappings L_1, L_2 such that $(L_1(a_0)L_1(a_1) \cdots L_1(a_n))$ and $(L_2(a_0)L_2(a_1) \cdots L_2(a_n))$ are in the same orientation of $\bar{y}_2^n$ and such that L_1 is positive and L_2 is negative. Now let P be the affine transformation which maps $L_1(a_i)$ into $L_2(a_i)$ for $i = 0, 1, \cdots, n$. It is easy to show from the theory of linear equations that such a transformation P exists. Since $(L_1(a_0) \cdots L_1(a_n))$ and $(L_2(a_0) \cdots L_2(a_n))$ are in the same orientation of $\bar{y}_2^n$, then it follows that P is positive. But

$$L_2 = PL_1$$

and thus the determinant of L_2 is the product of the determinants of P and L_1. Since P and L_1 are both positive, then PL_1 is positive. A contradiction. ∎

Definition. Let $\{L_\nu\}$ be the set of affine mappings of R^n into itself which map the oriented simplex y_1^n onto the oriented simplex y_2^n. If all the L_ν are positive, then y_1^n and y_2^n *have the same orientation.* If all the L_ν are negative, then y_1^n and y_2^n *have opposite orientations.*

From the properties of determinants and matrices, we obtain:

Lemma. *The relation of having the same orientation is an equivalence relation and this equivalence relation gives rise to two equivalence classes which we denote by* O_1 *and* O_2.

Definition. An *oriented* R^n is an R^n in which one of the equivalence classes, say O_1, has been designated as the class of positively oriented or positive simplexes and O_2 has been designated as the class of negatively oriented or negative simplexes. (Note that again the terms positive and negative are assigned in an arbitrary manner.)

Intersection numbers.

Definition. Let x^n be an oriented simplex in an oriented R^n, and let p be a point in R^n. If $p \notin \bar{x}^n$, then the *intersection number of* x^n *with* p, denoted by $i(x^n, p)$, is zero. If $p \in \bar{x}^n$, then $i(x^n, p)$ is $+1$ if x^n is a positive simplex in R^n and $i(x^n, p)$ is -1 if x^n is a negative simplex in R^n. The *intersection number of* p *with* x^n, denoted by $i(p, x^n)$, is $(-1)^n i(x^n, p)$.

Definition. Let H^{n-1} be a hyperplane in an oriented R^n (i.e., the set of points in R^n which satisfy a linear equation). H^{n-1} is a Euclidean $(n-1)$-space and we assume that H^{n-1} is oriented. Let H^1 be a line in R^n (i.e., the set of points in R^n which satisfy $n - 1$ independent linear equations).

H^1 is a Euclidean 1-space, and we assume that H^1 is oriented. Suppose the intersection of H^1 and H^{n-1} is a single point p. Let $(pa_1 \cdots a_{n-1})$ and (pb_1) be positively oriented simplexes in H^{n-1} and H^1 respectively. Then the *intersection number of* H^{n-1} *with* H^1, denoted by $i(H^{n-1}, H^1)$ is $+1$ $[-1]$ if the simplex $(pa_1 \cdots a_{n-1}b_1)$ is a positive [negative] simplex in R^n. The *intersection number of* H^1 *with* H^{n-1}, denoted by $i(H^1, H^{n-1})$ is $+1$ $[-1]$ if the simplex $(pb_1a_1 \cdots a_{n-1})$ is positive [negative] in R^n.

REMARK. In the definition we use the fact that $(pa_1 \cdots a_{n-1}b_1)$ is a simplex. This follows from the fact that H^1 and H^{n-1} have just one point, the point p, in common.

REMARK. To complete the definition, we must show that $i(H^{n-1}, H^1)$ and $i(H^1, H^{n-1})$ are independent of the choice of the points $a_1, \cdots, a_{n-1}$ and b_1. Let $(pa'_1 \cdots a'_{n-1}b'_1)$ and $(p\, b'_1)$ be positive simplexes in H^{n-1} and H^1 respectively. There is a positive affine mapping f from H^{n-1} into H^{n-1} such that $f(p) = p$ and $f(a_i) = a'_i$ $(i = 1, \cdots, n-1)$ and a positive mapping g from H^1 into H^1 such that $g(p) = p$ and $g(b_1) = b'_1$. Let F be the affine mapping of R^n into R^n such that $F/H^{n-1} = f$ and $F/H^1 = g$. Then F is a positive affine mapping. Hence $(pa'_1 \cdots a'_{n-1}b'_1)$ and $(pa_1 \cdots a_nb_1)$ are both positive or both negative.

LEMMA. $i(H^{n-1}, H^1) = (-1)^{n-1}\, i(H^1, H^{n-1})$.

PROOF. The proof follows from the definition of intersection number and orientation.

DEFINITION. Simplexes $\bar{x}^p$ and $\bar{y}^q$ in R^n are *in general position in* R^n if the union of the vertices in $\bar{x}^p$ and $\bar{y}^q$ has the property that any r points in it, where $r \leqq n + 1$, are linearly independent.

REMARK. If simplexes $\bar{x}^{n-1}$ and $\bar{y}^1$ in R^n are in general position in R^n then $\bar{x}^{n-1} \cap \bar{y}^1$ is a single point or the null set. For suppose $\bar{x}^{n-1} \cap \bar{y}^1 \neq \varnothing$ and contains more than one point. Then a line segment contained in $\bar{y}^1$ is contained in $\bar{x}^{n-1}$, and the vertices of $\bar{x}^{n-1}$ and one vertex (either vertex) of $\bar{y}^1$ are a set of $n + 1$ points which are not linearly independent.

DEFINITION. If $x^{n-1} = (a_0a_1 \cdots a_{n-1})$ and $y^1 = (b_0b_1)$ are oriented cells in an oriented R^n such that $\bar{x}^{n-1} \cap \bar{y}^1 = \varnothing$, then the *intersection number of* x^{n-1} *with* y^1, denoted by $i(x^{n-1}, y^1)$, and the *intersection number of* y^1 *with* x^{n-1}, denoted by $i(y^1, x^{n-1})$, are both zero. *If* x^{n-1} and y^1 are oriented simplexes and are in general position and if $\bar{x}^{n-1} \cap \bar{y}^1 \neq \varnothing$, then x^{n-1} and y^1 determine hyperplane H^{n-1} and line H^1, respectively, and $H^{n-1} \cap H^1$ is a single point. Suppose H^{n-1} and H^1 are oriented so that x^{n-1} is a positive simplex in H^{n-1} and y^1 a positive simplex in H^1. The *intersection number of* x^{n-1} *with* y^1, denoted by $i(x^n, y^1)$, is defined as:

$$i(x^{n-1}, y^1) = i(H^{n-1}, H^1).$$

The *intersection number of* y^1 *with* x^{n-1}, denoted by $i(y^1, x^{n-1})$, is defined as:

$$i(y^1, x^{n-1}) = i(H^1, H^{n-1}).$$

DEFINITION. Let

$$c^p = \sum_a u^a x_a^p,$$

$$d^q = \sum_b v^b y_b^q$$

be chains in R^n. The chains c^p and d^q are *in relatively general position* if for each pair of cells $\bar{x}_a^p$, $\bar{y}_b^q$, one of the following conditions holds:

(1) $\bar{x}_a^p \cap \bar{y}_b^q = \varnothing$;

(2) $\bar{x}_a^p$ and $\bar{y}_b^q$ are simplexes and are in general position.

DEFINITION. Suppose

$$c^p = \sum_a u^a x_a^p,$$

$$d^q = \sum_b v^b y_b^q,$$

where $p = n - 1$, $q = 1$, or $p = n$, $q = 0$, and c^p, d^q are chains in an oriented R^n which are in relatively general position. If neither c^p nor d^q is a null chain the *intersection number of* c^p *with* d^q, denoted by $i(c^p, d^q)$, is defined as:

$$i(c^p, d^q) = \sum_{a,b} i(x_a^p, y_b^q) u^a v^b$$

and the intersection number of d^q with c^p is

$$i(d^q, c^p) = \sum_{a,b} i(y_b^q, x_a^p) u^a v^b.$$

If c^p or d^q is a null chain, then $i(c^p, d^q) = i(d^q, c^p) = 0$.

The following statements stem directly from the definition:

(1) $i(c^p, d_1^q + d_2^q) = i(c^p, d_1^q) + i(c^p, d_2^q)$;

(2) $i(c_1^p + c_2^p, d^q) = i(c_1^p, d^q) + i(c_2^p, d^q)$;

(3) if a is an integer,

$$i(ac^p, d^q) = i(c^p, ad^q) = a[i(c^p, d^q)].$$

(2.2) THEOREM. *If the chains* c^n, d^1 *are in relatively general position, then*

$$i(c^n, b(d^1)) = (-1)^n i(b(c^n), d^1).$$

PROOF. It is sufficient to prove the statement for $c^n = x^n$, $d^1 = y^1$ where $\bar{x}^n \cap \bar{y}^1 \neq \varnothing$. Since $\bar{x}^n$ and $\bar{y}^1$ are in general position, then $\bar{x}^n \cap \bar{y}^1$ is a line segment $\overline{\alpha\beta}$ with endpoints α and β such that the line segment has at most one point in common with any $(n - 1)$-dimensional face of $\bar{x}^n$.

Case I. The line segment $\alpha\beta$ is contained in the interior of $\bar{x}^n$. (See Figure 3.) Then the line segment coincides with $\bar{y}^1$. Suppose y^1 has value $+1$ on the oriented simplex $(\alpha\beta)$, i.e., $y^1 = 1 \cdot (\alpha\beta)$. Since $\bar{y}^1$ is contained in the interior of $\bar{x}^n$, then

$$i(b(x^n), y^1) = 0.$$

But

$$i(x^n, b(y^1)) = i(x^n, \beta - \alpha) = i(x^n, \beta) - i(x^n, \alpha).$$

(In the expression $i(x^n, \beta)$, β denotes the 0-chain which has value 1 on β and is zero elsewhere.)
If x^n is a positive simplex in R^n, then $i(x^n, \beta) = i(x^n, \alpha) = 1$. If x^n is a negative simplex in R^n, then $i(x^n, \beta) = i(x^n, \alpha) = -1$. In either case, we obtain: $i(x^n, b(y^1)) = 0$.

Case II. $\overline{\alpha\beta}$ is contained in the interior of $\bar{y}^1$. Since $\bar{x}^n$ and $\bar{y}^1$ are in general position, then $\bar{y}^1$ intersects the interiors of two $(n-1)$-dimensional faces, say $\bar{x}_1^{n-1}$ and $\bar{x}_2^{n-1}$, of $\bar{x}^n$. Suppose $y^1 = (a_0a_1)$. Since $a_0 \notin \bar{x}^n$ and $a_1 \notin \bar{x}^n$, then

$$i(x^n, b(y^1)) = i(x^n, a_1 - a_0) = i(x^n, a_1) - i(x^n, a_0) = 0 - 0 = 0.$$

Now

$$i(b(x^n), y^1) = i(x_1^{n-1}, y^1) + i(x_2^{n-1}, y^1).$$

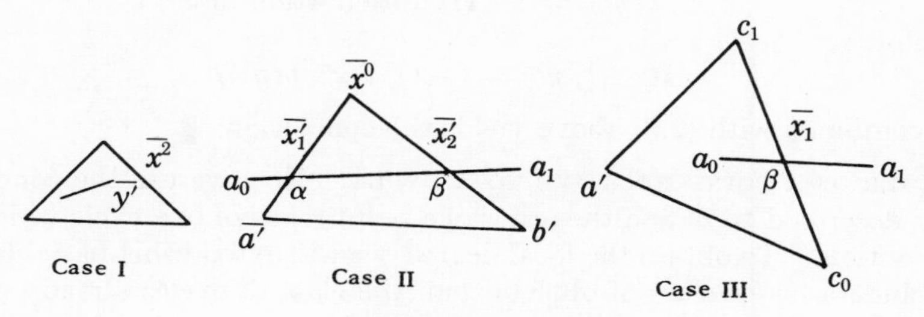

FIGURE 3

Let $\bar{x}^{n-2}$ be the common side of $\bar{x}_1^{n-1}$ and $\bar{x}_2^{n-1}$ and let a^1, b^1 be the vertices of $\bar{x}^n$ which are not in $\bar{x}^{n-2}$. The simplexes $(a^1 x^{n-2})$ and $(b^1 x^{n-2})$, where x^{n-2} is some orientation of $\bar{x}^{n-2}$, must appear in $b(x^n)$ with opposite signs because each contributes x^{n-2} to $b[b(x^n)]$ which is θ_{n-2} and x^{n-2} is not contributed to $b[b(x^n)]$ by any other term. Thus:

$$\begin{aligned} i(b(x^n), y^1) &= i(x_1^{n-1}, y^1) + i(x_2^{n-1}, y^1) \\ &= \pm[i(a^1 x^{n-2}, y^1) - i(b^1 x^{n-2}, y^1)] \\ &= \pm[i(\alpha x^{n-2}, y^1) - i(\beta x^{n-2}, y^1)] \\ &= \pm[\text{orientation of } (\alpha x^{n-2} a_1) \\ &\qquad - \text{orientation of } (\beta x^{n-2} a_1)] = 0. \end{aligned}$$

(Here "orientation of $(\alpha x^{n-2}a_1)$" means $+1$ or -1 depending whether simplex $(\alpha x^{n-2}a_1)$ is oriented positively or negatively.)

Case III. One endpoint, say a_0, of $\bar{y}^1$ is in the interior of $\bar{x}^n$ and the endpoint a_1 is such that $a_1 \notin \bar{x}^n$. Then $\bar{y}^1$ intersects just one $(n-1)$-dimensional side, $\bar{x}^{n-1}$, of $\bar{x}^n$, and the intersection is a single point β.

$$\begin{aligned} i(x^n, b(y^1)) &= i(x^n, a_1 - a_0) = i(x^n, a_1) - i(x^n, a_0) \\ &= 0 - i(x^n, a_0) = (-1) \text{ (orientation of } x^n). \end{aligned}$$

Suppose $x^{n-1} = (c_0 c_1 \cdots c_{n-1})$ where we choose the order in such a way that x^{n-1} has coefficient $+1$ in $b(x^n)$. Then

$$i(b(x^n), y^1) = i(x^{n-1}, y^1). \tag{2.3}$$

But

$$\begin{aligned} i(x^{n-1}, y^1) &= \text{orientation in } R^n \text{ of } (\beta c_1 \cdots c_{n-1}\, a_1) \\ &= (-1)^n[\text{orientation of } (a_1 \beta c_1 \cdots c_{n-1})] \\ &= (-1)^n[\text{orientation of } (a_1 c_0 c_1 \cdots c_{n-1})]. \end{aligned}$$

Let a^1 be the vertex of $\bar{x}^n$ that is not a vertex of $\bar{x}^{n-1}$. Then a^1 is on the other side of $\bar{x}^{n-1}$ from a_1. The orientation of $(a_1 c_0 c_1 \cdots c_{n-1})$ is the opposite of that of x^n. Therefore

$$i(x^{n-1}, y^1) = (-1)^{n+1} \quad \text{(orientation of } x^n).$$

But

$$i(x^n, b(y^1)) = (-1) \quad \text{(orientation of } x^n).$$

Therefore

$$i(x^{n-1}, y^1) = (-1)^n\, i(x^n, b(y^1)).$$

This combined with (2.3) above yields the conclusion. ∎

ORDER OF A POINT RELATIVE TO A CYCLE. Now we use the concepts already described to define the order of a point relative to a cycle which is our basic notion. To obtain the local degree, we will extend this basic definition to include a wider class of objects than the class of cycles already defined.

NOTATION. Let $\bar{c}^n$ denote the union of the cells on which the chain c^n has nonzero values.

(2.4) LEMMA. *If z^n is a cycle in R^n and $p \in R^n$, then*

$$i(z^n, p) = 0.$$

PROOF. Let $q \in R^n$ be such that $q \notin \bar{z}^n$. Then $i(z^n, q) = 0$. Let $c^1 = (qp)$. Then $b(c^1) = p - q$. Then

$$\begin{aligned} i(z^n, p - q) &= i(z^n, p) - i(z^n, q) \\ &= i(z^n, b(q\,p)) \\ &= (-1)^n\, i(b(z^n), (qp)) \\ &= (-1)^n\, i(\theta_{n-1}, (qp)) \\ &= 0. \end{aligned}$$

Thus $i(z^n, p) = i(z^n, q) = 0$. ∎

Now let z^{n-1} be a cycle in R^n (where $n \geqq 2$) and let p be a point such that $p \notin z^{n-1}$. By Theorem (2.1) there is a chain c^n such that

$$b(c^n) = z^{n-1}.$$

Also c^n may be constructed so that p and c^n are in relatively general position. If c_1^n and c_2^n are two such chains, then

$$b(c_1^n - c_2^n) = z^{n-1} - z^{n-1} = \theta_{n-1}.$$

That is, the chain $c_1^n - c_2^n$ is a cycle and hence by Lemma (2.4),

$$i(c_1^n - c_2^n, p) = 0$$

or

$$i(c_1^n, p) = i(c_2^n, p).$$

Thus the following definition is justified.

(2.5) DEFINITION. If z^{n-1} is a cycle in R^n with $n \geqq 2$ and p is a point such that $p \notin \bar{z}^{n-1}$, let c^n be a chain such that $b(c^n) = z^{n-1}$ and such that c^n and p are in relatively general position. The *order of p relative to z^{n-1}*, denoted by $v(z^{n-1}, p)$, is defined as:

$$v(z^{n-1}, p) = i(c^n, p).$$

REMARK. An immediate consequence of this definition is: if $v(z^{n-1}, p) \neq 0$ and $b(c^n) = z^{n-1}$, then $p \in \bar{c}^n$. As we shall see, this simple observation is the basis for all applications in analysis of local degree.

REMARK. From the definition, it follows that

$$v(az_1^{n-1} + bz_2^{n-1}, p) = av(z_1^{n-1}, p) + bv(z_2^{n-1}, p).$$

(2.6) THEOREM. *If $z_1^{n-1} \sim z_2^{n-1}$ in the open set $R^n - p$, then*

$$v(z_1^{n-1}, p) = v(z_2^{n-1}, p).$$

PROOF. By hypothesis, there is a chain c^n on $R^n - p$ such that

$$b(c^n) = z_1^{n-1} - z_2^{n-1}.$$

Therefore

$$v(z_1^{n-1} - z_2^{n-1}, p) = i(c^n, p).$$

But c^n is on $R^n - p$, i.e., $p \notin \bar{c}^n$. Hence $i(c^n, p) = 0$, and thus

$$v(z_1^{n-1} - z_2^{n-1}, p) = 0$$

or

$$v(z_1^{n-1}, p) = v(z_2^{n-1}, p). \blacksquare$$

Suppose $p \notin \bar{z}^{n-1}$ where z^{n-1} is a cycle in R^n. Let c^n be a chain such that $b(c^n) = z^{n-1}$ and such that c^n and p are in relatively general position. Let N be a sphere in R^n with center at the origin and such that $\bar{c}^n \subset N$. Let q

be a point such that $q \notin N$ and such that the line segment $\overline{pq}$ intersects $\bar{z}^{n-1}$ in at most a finite number of points. (Since $p \notin \bar{z}^{n-1}$, it is easy to show that such a point q exists.) Then (see Figure 4)

$$\begin{aligned} i((pq), z^{n-1}) &= i((pq), b(c^n)) \\ &= -i(q - p, c^n), \text{ by Theorem (2.2),} \\ &= -i(q, c^n) + i(p, c^n). \end{aligned}$$

Since $q \notin \bar{c}^n$, then $i(q, c^n) = 0$. Thus we obtain

(2.7) THEOREM. $i((pq), z^{n-1}) = i(p, c^n)$.

This formula yields a convenient method for computing $v(z^{n-1}, p)$. Indeed one of our main purposes in choosing this approach to the concept of local degree is to be able to utilize this formula.

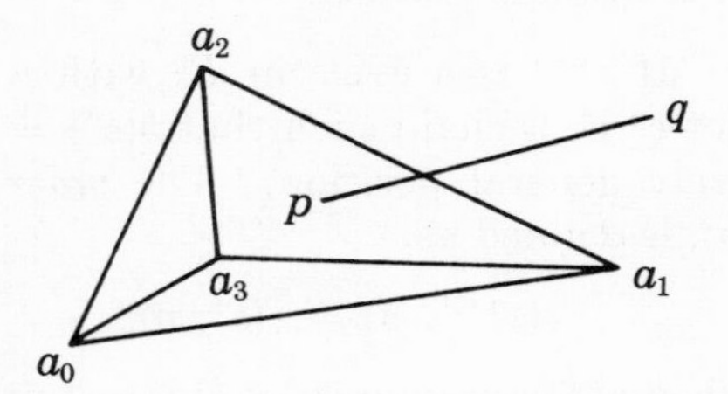

$$c^2 = (a_0a_1a_3) + (a_1a_2a_3) + (a_3a_2a_0).$$
$$b(c^2) = (a_0a_1) + (a_1a_2) + (a_2a_0).$$

FIGURE 4

3. **The order of a point relative to a continuous image of $\bar{z}^{n-1}$.** The purpose of this section is to extend the concept $v(z^{n-1}, p)$ to the case in which z^{n-1} is replaced by a continuous image of $\bar{z}^{n-1}$. If $f(\bar{z}^{n-1})$ is such a continuous image, the idea is to approximate $f(\bar{z}^{n-1})$ by a cycle z_1^{n-1} and then compute $v(z_1^{n-1}, p)$. To show that this idea can be used, we must show that any sufficiently good approximation z_1^{n-1} can be used, i.e., that all such approximations yield the same result. To prove this, we must introduce some more formal concepts. The reader who wishes may omit this section except for the definition of simplicial complex immediately below and go on at once to Definition (3.13a).

COMPLEXES: SUBDIVISIONS, CHAINS, AND CHAIN MAPPINGS.

DEFINITION. A *cell complex* $\mathscr{K}$ is a finite collection of cells such that:

(1) if $\bar{x}^k \in \mathscr{K}$ and $\bar{y}^l$ is a side of $\bar{x}^k$, then $\bar{y}^l \in \mathscr{K}$;

(2) if $\bar{x}^k, \bar{y}^l \in \mathscr{K}$, then $\bar{x}^k \cap \bar{y}^l$ is the null set or a cell $\bar{x}^q$ such that $\bar{x}^q$ is a side of $\bar{x}^k$ and a side of $\bar{y}^l$.

DEFINITION. A *simplicial complex* K is a cell complex in which all the cells are simplexes.

DEFINITION. An *n-dimensional cell complex* is a complex which is a finite collection of n-cells and their sides.

DEFINITION. A *polyhedron* P is the union of the simplexes in a simplicial complex K. The polyhedron P is also denoted by $\overline{K}$. The simplicial complex K is called a *simplicial decomposition* of polyhedron P.

NOTATION. Let $\overline{\mathscr{K}}$ be the union of all the cells in $\mathscr{K}$, and $\overline{K}$ the union of the cells in K.

DEFINITION. A simplicial complex K is a *simplicial subdivision* of cell complex $\mathscr{K}$ if:

(i) each simplex of K is a subset of a cell of $\mathscr{K}$;

(ii) $\overline{K} = \overline{\mathscr{K}}$.

(See Figure 5)

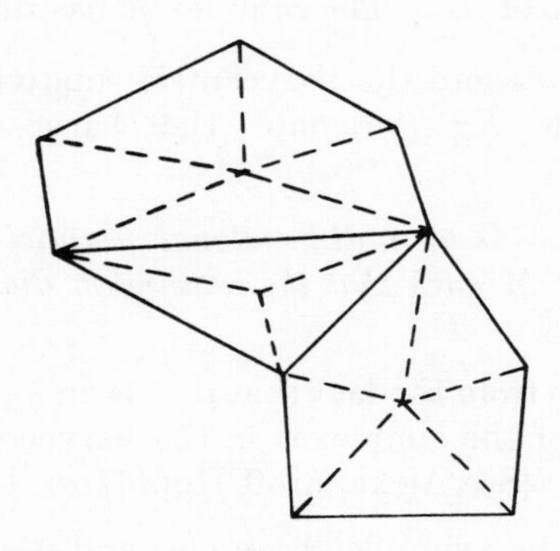

FIGURE 5. A SIMPLICIAL SUBDIVISION

(3.1) THEOREM. *If $\mathscr{K}$ is an n-dimensional cell complex, there exists a simplicial subdivision K of $\mathscr{K}$.*

PROOF. The proof is by induction on the dimension of $\mathscr{K}$. If $\mathscr{K}$ is 0-dimensional, then $\mathscr{K}$ itself is a simplicial subdivision. Now suppose the theorem is true if $\mathscr{K}$ is of dimension $m - 1$. Consider $\mathscr{K}$, an m-dimensional cell complex. The collection of sides of the m-cells in $\mathscr{K}$ is an $(m - 1)$-dimensional cell complex $\mathscr{K}_1$. By the induction assumption there is a simplicial complex K_1 which is a simplicial subdivision of $\mathscr{K}_1$. Let $\bar{c}_1^n, \cdots, \bar{c}_q^n$ be the collection of n-cells in $\mathscr{K}$. The point set boundary of $\bar{c}_i^n$ $(i=1, \cdots, q)$ is the union of ν $(n - 1)$-simplexes

$$\bar{x}_{ij}^{n-1} = \overline{(a_{ij0}a_{ij1}\cdots a_{ijn-1})} \qquad (j = 1, \cdots, \nu).$$

Let o_i be a point in the interior of $\bar{c}_i^n$. It is straightforward to verify that the desired K is the collection of n-simplexes:

$$(o_i a_{ij0} a_{ij1} \cdots a_{ijn-1}) \qquad (i = 1, \cdots, q;\ j = 1, \cdots, \nu)$$

and their sides. ▌

DEFINITION. If the cell complex $\mathscr{K}$ of Theorem (3.1) is a simplicial complex and each point o_i is the center of gravity of the simplex in whose interior o_i is contained, then the simplicial subdivision described in the proof is the *barycentric subdivision* of $\mathscr{K}$.

(3.2) THEOREM. *If $\mathscr{K}_1$, $\mathscr{K}_2$ are two n-dimensional cell complexes such that $\overline{\mathscr{K}}_1 = \overline{\mathscr{K}}_2$, then there is an n-dimensional simplicial complex K such that K is a simplicial subdivision of $\mathscr{K}_1$ and a simplicial subdivision of $\mathscr{K}_2$.*

PROOF. The collection of cells which are the intersections of the cells in $\mathscr{K}_1$ and $\mathscr{K}_2$ is a cell complex $\mathscr{K}$. By the preceding theorem there is a simplicial subdivision K of $\mathscr{K}$. The complex K has the desired properties. ▌

NOTATION. Let K_1^B denote the barycentric subdivision of the simplicial complex K and let K_{m+1}^B denote the barycentric subdivision of K_m^B $(m = 1, 2, \cdots)$.

(3.3) THEOREM. *If K is an n-dimensional simplicial complex and $\varepsilon > 0$, then there is an integer M such that the maximum diameter of a simplex in K_M^B is $< \varepsilon$.*

PROOF. This follows from the fact that if $\bar{x}^n$ is an n-simplex with diameter d then the diameters of the simplexes in the barycentric subdivision of $\bar{x}^n$ are all $\leqq (n/(n+1))d$. (See Alexandroff-Hopf [**1**, p. 136].)

DEFINITION. Let K be a simplicial complex and assume that each simplex in K has been assigned an orientation. As before, we use the notations x^q and $-x^q$. Then K is said to be an *oriented simplicial complex*.

DEFINITION. Let the oriented simplicial complex K be such that $\bar{K} \subset R^n$. If $q \leqq n$, a *q-chain on K* is a q-chain on R^n which is nonzero only on oriented q-simplexes x^q such that $x^q \subset \bar{K}$ and the oriented q-simplexes $-x^q$.

DEFINITION. Let K_1 denote a simplicial subdivision of a simplicial complex K where $\bar{K} \subset R^n$. For $m \geqq 1$, let K_{m+1} denote a simplicial subdivision of K_m $(m = 1, 2, \cdots)$. If K is oriented, let K_{m+1} denote the oriented simplicial complex defined this way: if $\bar{x}_{m+1}^q$ is a q-simplex in K_{m+1}, then $\bar{x}_{m+1}^q \subset \bar{x}^q$ where x^q is an oriented simplex in K. Suppose

$$\bar{x}_{m+1}^q = \overline{(b_0 b_1 \cdots b_q)}$$

and

$$x^q = (a_0 a_1 \cdots a_q).$$

Let L be the affine mapping such that $L(a_i) = b_i$ for $i = 0, 1, \cdots, q$. If L is positive, we define x^q_{m+1} to be the oriented q-simplex

$$(L(a_0)L(a_1)\cdots L(a_q)) = (b_0b_1\cdots b_q).$$

If L is negative, we define x^q_{m+1} to be the oriented q-simplex $(b_1b_0\cdots b_q)$. (From the intuitive viewpoint, we are merely saying: if x^q is oriented clockwise, then x^q_{m+1} is also to be oriented clockwise.)

REMARK. From now on, the term complex will mean always an oriented q-dimensional simplicial complex and subdivisions of the complex will be oriented as described in the preceding definition.

(3.4) DEFINITION. Let $\sum t^i x^q_i$ be a q-chain on complex K. Suppose $x_{i1}, \cdots, x_{im_i}$ are the q-simplexes in K_m such that $\bar{x}^q_i = \bigcup_{j=1}^{m_i} \bar{x}_{ij}$. The *chain mapping S_m induced by m subdivisions* is the mapping:

$$S_m : \sum_i t^i x^q_i \to \sum_i (t^i x_{i1} + \cdots + t^i x_{im_i}).$$

THEOREM. $bS_m = S_m b$.

PROOF. It is sufficient to prove the statement for the chain x^q which is 1 on the oriented q-simplex x^q in K and zero on all other q-simplexes x^q such that $\bar{x}^q \subset \bar{K}$. First by straightforward verification, we have:

$$bS_1(x^q) = S_1 b(x^q). \tag{3.5}$$

Next assume that

$$bS_{n-1}(x^q) = S_{n-1}b(x^q).$$

Let σ_1 denote the chain mapping induced by the single subdivision K_m of K_{m-1}. Then

$$\begin{aligned} bS_m(x^q) &= b\sigma_1 S_{m-1}(x^q) \quad \text{by definition of } \sigma_1 \text{ and } S_{m-1}; \\ &= \sigma_1 b S_{m-1}(x^q) \quad \text{by (3.5)}; \\ &= \sigma_1 S_{m-1} b(x^q) \quad \text{by the induction assumption}; \\ &= S_m b(x^q) \quad \text{by the definition of } \sigma_1 \text{ and } S_{m-1}. \ \blacksquare \end{aligned}$$

(3.6) COROLLARY. *If z^q is a cycle on K, then $S_m(z^q)$ is a cycle on K_m.*

DEFINITION. A *simplicial mapping* f of a complex K into an open set $U \subset R^n$ is a mapping from the vertices of K into U.

DEFINITION. The *chain mapping F (induced by a simplicial mapping f)* of the chains on K into the chains on U is the mapping defined as follows:

(1) let $(a_0a_1\cdots a_n)$ be an oriented simplex in K. If the points $f(a_0)$, $f(a_1), \cdots, f(a_n)$ are all distinct, then

$$F[(a_0a_1\cdots a_n)] = (f(a_0)f(a_1)\cdots f(a_n));$$

(2) if the points $f(a_0), \cdots, f(a_n)$ are not all distinct, then

$$F[(a_0a_1 \cdots a_n)] = \theta_n;$$

(3) if $\sum t^i x_i$ is a chain on K, then

$$F[\sum t^i x_i] = \sum t^i F[x_i].$$

A straightforward verification proves:

(3.7) THEOREM. *$bF = Fb$ for all chains on K.*

(3.8) COROLLARY. *If z is a cycle on K, then $F(z)$ is a cycle in U.*

(3.8a) DEFINITION. Let f be a simplicial mapping of K into U. If $x^n \in K$, then let f_1 denote f on the vertices of x^n. (More precisely, if $\bar{x}^n = \overline{(a_0a_1 \cdots a_n)}$, then $f_1 = f / \bigcup_{i=0}^n a_i$.) Extend f_1 linearly over $\bar{x}^n$. That is, if $\underline{(a_0a_1 \cdots a_q)}$ is a side of $\bar{x}_n$ and if p is a point in the interior of $\overline{a_0a_1 \cdots a_q}$, then p may be written as a linear combination of $a_0, a_1, \cdots, a_q$,

$$p = \sum_{i=0}^{n} \gamma_i a_i$$

and we define

$$f_L(p) = \sum_{i=0}^{n} \gamma_i f_L(a_i).$$

Applying this procedure to each $x^n \in K$, we obtain a continuous mapping f_L of $\bar{K}$ into R^n which is linear on each $\bar{x}^n$. The mapping f_L is called the *continuous simplicial mapping* of $\bar{K}$ into R^n induced by the simplicial mapping f.

Now let K be a complex such that $\bar{K} \subset R^n$ and regard R^n as a subset of R^{n+1}. Let

$$P = \bigcup_{p \in \bar{K}} L(p)$$

where $L(p)$ is a line segment of unit length such that p is one endpoint of $L(p)$, and $L(p)$ is perpendicular to R^n, and all the line segments $L(p)$ are on the same side of R^n in R^{n+1}. Let p' denote the other endpoint of $L(p)$. Now if $(a_0a_1 \cdots a_q)$ is a simplex in K, the collection of cells of the form $\overline{(a_0a_1 \cdots a_q a_0' a_1' \cdots a_q')}$ is a cell complex K. We obtain a simplicial subdivision $Z(K)$ of K as follows: suppose $x_i \in K$ and let

$$Z(\bar{x}_i) = \bigcup_{p \in \bar{x}_i} L(p).$$

Let o_i be a fixed point (called the center) in the interior of $Z(\bar{x}_i)$. Then if $\bar{x}_i$ is a 0-simplex, the line $Z(\bar{x}_i)$ is divided by o_i into two 1-simplexes. Now suppose that the subdivision has been constructed for all the cells of the form $Z(\bar{x}_i^j)$ where $j \leqq q - 1$. Then the boundary of the cell

$$S = \overline{(a_0a_1 \cdots a_q a_0' a_1' \cdots a_q')}$$

is the union of q-simplexes already constructed. Let p be the center of S. Using the same construction as in the proof of Theorem (3.1), we obtain a simplicial subdivision, denoted by $Z(K)$, of K.

We assign orientations to the simplexes of $Z(K)$ this way: if $(a_0a_1\cdots a_q)$ is a q-simplex in K, the simplex $\overline{(a_0a_1\cdots a_q)}$ in $Z(K)$ is assigned the orientation $(a_0a_1\cdots a_q)$, i.e., the simplex has the same orientation that it has in K. Let the orientation of $\overline{(a_0'a_1'\cdots a_q')}$ be $(a_0'a_1'\cdots a_q')$. Finally we assign arbitrary orientations to the remaining simplexes of $Z(K)$.

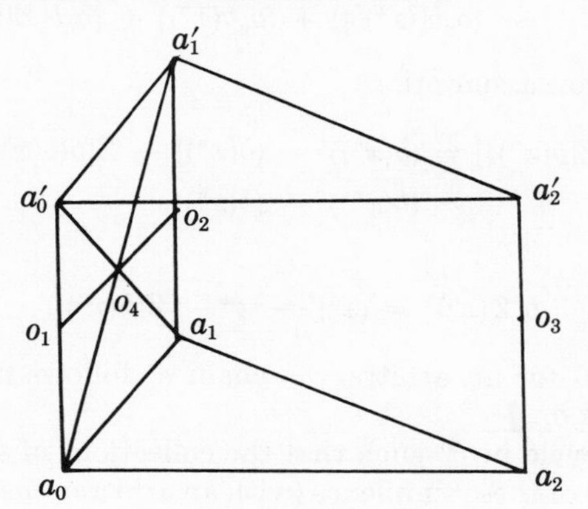

FIGURE 6. CONSTRUCTION OF $Z(K)$

DEFINITION. The complex $Z(K)$ is called a *cylinder over the complex* K.

Now if the complex K consists of simplexes $x_m^q = (a_0a_1\cdots a_q)$ $(m = 1,\cdots,w)$, let K' denote the complex which consists of the simplexes $(x_m^q)' = (a_0'a_1'\cdots a_q')$. If $c = \sum_a t^a x_a$ is a chain on K, let c' be the chain $\sum_a t^a x_a'$.

(3.9) THEOREM. *If z is a cycle on K, then $z \sim z'$ on $Z(K)$.*

PROOF. First corresponding to each chain c^p on K we define a chain $Z(c^p)$ in $Z(K)$. If c^0 is a 0-chain in K, define $Z(c^0) = \theta_1$. Now assume $Z(c^{r-1})$ has been defined for all $(r-1)$-chains in K. Let x^r be the chain on K which has value $+1$ on simplex x^r and is zero elsewhere. Define

$$Z(x^r) = (o_r(x^r)') - (o_r x^r) - (o_r Z[b(x^r)])$$

where o_r is the center of the cell $(x^r(x^r)')$, i.e., the cell whose vertices are those of x^r and those of $(x^r)'$, described in the definition of $Z(K)$, and we use the same kind of notation as in the proof of Theorem (2.1). If $c^r = \sum t^i x_i$, define

$$Z(c^r) = \sum t^i Z(x_i).$$

If $c^r = \theta_r$, then $Z(c^r) = \theta_{r+1}$.

We complete the proof by showing that

$$b[Z(c^r)] = (c^r)' - c^r - Z[b(c^r)]. \tag{3.10}$$

Statement (3.10) completes the proof because if c^r is a cycle, then $b(c^r) = \theta_{r-1}$ and $Z(\theta_{r-1}) = \theta_r$.

We prove (3.10) by induction. The proof of (3.10) for the case $r = 1$ is a straightforward verification. Now suppose (3.10) holds for $r = n - 1$. Then:

$$\begin{aligned} b[Z(x^n)] &= b\{(o_n(x^n)') - (o_n x^n) - (o_n Z[b(x^n)])\} \\ &= (x^n)' - x^n - Z[b(x^n)] \\ &\quad - (o_n b[(x^n)']) + (o_n b(x^n)) + (o_n b[Z[b(x^n)]]). \end{aligned}$$

But by the induction assumption:

$$\begin{aligned} b[Z[b(x^n)]] &= (b(x^n))' - (b(x^n)) - Z[b(b(x^n))] \\ &= (b(x^n))' - (b(x^n)). \end{aligned}$$

Thus

$$b[Z(x^n)] = (x^n)' - x^n - Z[b(x^n)].$$

The proof of (3.10) for an arbitrary n-chain c^n follows from the additivity of operators Z and b. ▌

Now let z be a cycle in R^n such that the collections of simplexes on which $z \neq 0$ and the sides of these simplexes (with an arbitrary assigned orientation) are a complex K. Let f_1, f_2 be simplicial mappings from K into an open set $U \subset R^n$. Assume that for each simplex $x^q \in K$, there is a convex subset $Q \subset U$ such that $[f_{1L}(\bar{x}^q)] \cup [f_{2L}(\bar{x}^q)] \subset Q$ where f_{1L}, f_{2L} are the continuous simplicial mappings induced by f_1, f_2. Define a simplicial mapping f from $Z(K)$ into U as follows:

(1) if e is a vertex of a simplex of K, let $f(e) = f_1(e)$ and let $f(e') = f_2(e)$;

(2) if o_i is the center of $\overline{a_0 \cdots a_q a_0' \cdots a_q'}$, let $f(o_i)$ be an arbitrary point in the convex hull of $[f_{1L}(\overline{a_0 \cdots a_n})] \cup [f_{2L}(\overline{a_0' \cdots a_n'})]$.

Let F_1, F_2, F be the chain mappings induced by f_1, f_2, and f.

THEOREM. *$F_1(z) \sim F_2(z)$ in U.*

PROOF. Since $z \sim z'$ by Theorem (3.9), there is a chain c on $Z(K)$ such that $b(c) = z - z'$. But by Theorem (3.7),

$$bF(c) = Fb(c) = F(z - z') = F(z) - F(z').$$

From the definition of F_1, F_2, and F, we have:

$$F(z) = F_1(z)$$

and

$$F(z') = F_2(z). \ ▌$$

(3.11) COROLLARY. *If f_1, f_2 are simplicial mappings from the complex K described above into an open set $U \subset R^n$ such that if e is a vertex of a simplex of K, then*

$$|f_1(e) - f_2(e)| < \alpha = \tfrac{1}{2}\, d(f_{1L}(\bar{K}),\ R^n - U);$$

then if F_1, F_2 are the chain mappings induced by f_1, f_2, it is true that $F_1(z) \sim F_2(z)$ in U.

PROOF. Let $N_\alpha(p)$ be the α-neighborhood of a point p, i.e.,

$$N_\alpha(p) = [q/|p - q| < \alpha].$$

The set

$$Q = \bigcup_{p \in f_1(\bar{x}_i)} N_\alpha(p)$$

is a convex set and for each $x_i \in K$,

$$[f_{1L}(\bar{x}_i)] \cup [f_{2L}(\bar{x}_i)] \subset Q. \blacksquare$$

DEFINITION. Let ϕ be a continuous mapping of a polyhedron $P = \bar{K}$ into R^n. An *ε-approximation of ϕ* is a simplicial mapping f of a simplicial subdivision K_1 of the simplicial complex K into R^n such that the corresponding f_L has the property that if $x \in P$, then

$$|\phi(x) - f_L(x)| < \varepsilon.$$

(3.11a) THEOREM. *If ϕ is a continuous mapping of a polyhedron P into R^n, then for arbitrary $\varepsilon > 0$, there is an ε-approximation f of ϕ.*

PROOF. Since P is a closed bounded point set, the mapping ϕ is uniformly continuous on $P = \bar{K}$. Hence there is a $\delta > 0$ such that if $p_1, p_2 \in P$ and $|p_1 - p_2| < \delta$, then $|\phi(p_1) - \phi(p_2)| < \varepsilon/2$. By Theorem (3.3) there is an N such that the diameter of a simplex in K_N^B is less than δ. For each vertex e of a simplex of K_N^B, let $f(e) = \phi(e)$. It follows easily that f is an ε-approximation of ϕ. $\blacksquare$

(3.12) THEOREM. *Let K be an oriented $(n - 1)$-dimensional complex consisting of oriented $(n - 1)$-simplexes $x_1^{n-1}, \cdots, x_N^{n-1}$ and their sides and such that $b(\sum_{i=1}^N x_i^{n-1}) = \theta_{n-2}$. Let ϕ be a continuous mapping from $P = \bar{K}$ into U, an open subset of R^n. Let the point p be in $R^n - U$. Then there is a positive ε such that for every ε-approximation f, the corresponding chain mapping F maps into chains on U. Also if f_1, f_2 are two ε-approximations, of ϕ, and K_1, K_2 are the corresponding complexes and F_1, F_2 are the corresponding chain mappings, then*

$$v[F_1S_1(\textstyle\sum x_i^{n-1}), p] = v[F_2S_2\,(\sum x_i^{n-1}), p],$$

where S_1, S_2 are the chain mappings described in Definition (3.4).

PROOF. Let $\varepsilon = \frac{1}{3}d[\phi(\bar{K}), R^n - U]$. First by Corollaries (3.6) and (3.8), $F_i S_i(\sum_{i=1}^N x_i^{n-1})$ is a cycle. By Theorem (3.2), there is a complex K_3 such that K_3 is a simplicial subdivision of K_1, K_2, and K. Let S_{13} and S_{23} be the chain mappings from the chains of K_1 and K_2 into K_3. By Corollary (3.6) the chains $S_{13}S_1(\sum x_i^{n-1})$ and $S_{23}S_2(\sum x_i^{n-1})$ on K_3 are cycles. The simplicial mappings $f_{iL/[\text{set of vertices of } K_3]}$, where $i = 1, 2$, induce two chain mappings:

$$F_3^{(i)}: [\text{Chains on } K_3] \to [\text{Chains on } U]\ ,\ i=1,\ 2.$$

We complete the proof of the theorem by proving the following three statements:

(a) $v[F_1S_1(\sum x_i^{n-1}), p] = v[F_3^{(1)}S_{13}S_1(\sum x_i^{n-1}), p]$;
(b) $v[F_2S_2(\sum x_i^{n-1}), p] = v[F_3^{(2)}S_{23}S_2(\sum x_i^{n-1}), p]$;
(c) $v[F_3^{(1)}S_{13}S_1(\sum x_i^{n-1}), p] = v[F_3^{(2)}S_{23}S_2(\sum x_i^{n-1}), p]$.

By Corollary (3.11)

$$F_3^{(1)}S_{13}S_1(\sum x_i^{n-1}) \tilde{\sim} F_3^{(2)}S_{23}S_2(\sum x_i^{n-1})$$

in U. Hence by Theorem (2.6), equality (c) holds. Since (a) and (b) are similar, it is only necessary to prove (a). But (a) follows from Theorem (2.7) because $S_{13}S_1(\sum x_i^{n-1})$ is a chain on a complex which is a simplicial subdivision of the complex on which $S_1(\sum x_i^{n-1})$ is a chain. ∎

Now let ϕ, K satisfy the hypotheses of the preceding theorem and let $p \in R^n - U$ as before. The preceding Theorem (3.12) justifies the following definition:

(3.13) DEFINITION. Let f be an ε-approximation of ϕ where

$$\varepsilon < \tfrac{1}{3}\, d[\phi(\bar{K}), R^n - U].$$

The *order of p relative to ϕ and $\bar{K}$*, denoted by $v[\phi, \bar{K}, p]$, is the value $v(F(\sum x_i^{n-1}), p)$ where F is the chain mapping induced by f.

For the reader who has omitted §3, we repeat this definition in less precise intuitive language.

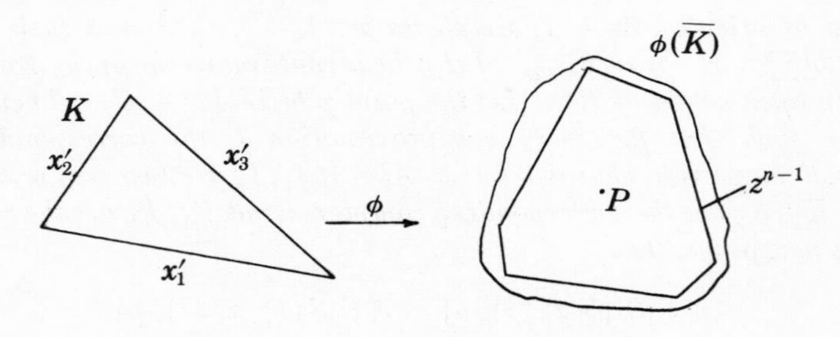

FIGURE 7. DEFINITION OF ORDER

(3.13a) INTUITIVE VERSION OF THE DEFINITION. Let K be a complex consisting of oriented $(n - 1)$-simplexes $x_1^{n-1}, \cdots, x_w^{n-1}$ and their sides (assigned an arbitrary orientation) such that the chain $\sum_{i=1}^{w} x_i^{n-1}$ is a cycle and suppose ϕ is a continuous mapping defined on $\overline{K}$ such that $\phi(\overline{K}) \subset U$, an open set in R^n. Let p be a point such that $p \in R^n - U$. Then if z^{n-1} is an $(n - 1)$-cycle such that the set $\bar{z}^{n-1}$ can be deformed continuously into $\phi(\overline{K})$ without passing through the point p, we define the *order of p relative to ϕ and $\overline{K}$*, denoted by $v(\phi, \overline{K}, p)$, to be $v(z^{n-1}, p)$. (See Figure 7.)

4. Properties of $v[\phi, \overline{K}, p]$.

DEFINITION. A *homotopy* is a continuous mapping from $\overline{K} \times [0, 1]$ into R^n. We denote the homotopy by $\phi_t : \overline{K} \to R^n$, i.e., subscript t denotes the point in $[0, 1]$.

NOTATION. Let $p(t)$ denote the image under a continuous mapping p from $[0, 1]$ into R^n.

(4.1) THEOREM OF INVARIANCE UNDER HOMOTOPY. *Let K denote the same type of complex as in Definition* (3.13) *and Theorem* (3.12). *Let ϕ_t be a homotopy and $p(t)$ the image under a continuous mapping such that for each $t_0 \in [0, 1]$, the point $p(t_0) \notin \phi_{t_0}(\overline{K})$. Then for all $t \in [0, 1]$, the order $v(\phi_t, \overline{K}, p(t))$ has the same value.*

PROOF. Let $t_0 \in [0, 1]$. Since $p(t_0) \notin \phi_{t_0}(\overline{K})$, there is an open set U such that $p(t_0) \notin U$ and $\phi_{t_0}(\overline{K}) \subset U$. Now $\varepsilon > 0$ implies there is a $\delta > 0$ such that if f is an $(\varepsilon/2)$-approximation of ϕ_{t_0} then for t such that $|t - t_0| < \delta$, the simplicial mapping f is an ε-approximation of ϕ_t. By this statement and the fact that $p(t)$ is the image of a continuous mapping, it follows that there is a $\delta_1 > 0$ and a pair of disjoint open sets N and $U_1 = U$ such that if $|t - t_0| < \delta_1$, then $p(t) \in N$ and $\phi_t(\overline{K}) \subset U_1$ and for all t such that $|t - t_0| < \delta$, the order $v(\phi_t, K, p(t))$ has the same value. For each $t_0 \in [0, 1]$, there is such a δ_1-neighborhood. The proof is completed by applying the Heine-Borel Theorem. ∎

(4.2) COROLLARY (POINCARÉ-BOHL THEOREM). *Let K be the same complex as before, and let ϕ_1, ϕ_2 be continuous mappings from $\overline{K}$ into R^n. Suppose further that a point p is given such that if $q \in \overline{K}$, then p is not on the line segment with endpoints $\phi_1(q)$ and $\phi_2(q)$. Then*

$$v(\phi_1, \overline{K}, p) = v(\phi_2, \overline{K}, p).$$

PROOF. For each $t \in [0, 1]$ and $q \in \overline{K}$, let $\phi_t(q)$ be the point in R^n which divides the line segment $\overline{\phi_1(q)\phi_2(q)}$ in the ratio $t/(1 - t)$ and apply Theorem (4.1). ∎

(4.3) THEOREM ON EXISTENCE OF A SOLUTION. *Let K be an n-dimensional complex and z^{n-1} a cycle on K. Suppose there is a chain c^n on K such that*

$$b(c^n) = z^{n-1}.$$

Suppose ϕ is a continuous mapping with domain $\bar{c}^n$ and range a subset of R^n. (Note that ϕ is defined on the subset $\bar{z}^{n-1}$ of $\bar{c}^n$.) Let p be a point such that $p \notin \phi(\bar{z}^{n-1})$. Suppose $v(\phi, \bar{z}^{n-1}, p) \neq 0$. Then $p \in \phi(\bar{c}^n)$.

PROOF. Let ϕ_1 be an ε-approximation of $\phi/\bar{z}^{n-1}$. If ε is sufficiently small, then $v(\Phi_1, (z^{n-1}), p) \neq 0$ where Φ_1 is the chain mapping induced by ϕ_1. By the remark following Definition (2.5), we have: $p \in \overline{\Phi_1(c^n)}$. Since ε can be made arbitrarily small and since $\phi(\bar{c}^n)$ is closed and bounded, then $p \in \phi(\bar{c}^n)$. ▮

5. **The local degree relative to a complex.** From one point of view, our definition of topological degree is now practically complete. If K is an n-dimensional complex consisting of n-simplexes $x_1^n, \cdots, x_w^n$, all with positive orientation in oriented R^n and if ϕ is a continuous mapping defined on $\bar{K}$ and p is a point such that $p \notin \phi(\bar{z}^{n-1})$, where $z^{n-1} = b(\sum_{i=1}^{w} x_i)$, then we define the local degree of ϕ at p relative to $\bar{K}$ to be

$$v\left[\phi, \overline{b\left(\sum_{i=1}^{w} x_i^n\right)}, p\right].$$

However, this places two undesirable limitations on the local degree. First we have not considered the degree as a sum of "indices" of points or as a "covering number." As we will see later, this is an important aspect of the theory because it suggests some methods for computing the degree. Also this aspect of the theory leads to estimates on how many points in $\bar{K}$ are mapped by ϕ onto the point p. Secondly we wish to extend our definition so that we may speak of the local degree of ϕ relative to a set E which is not of the form $\bar{K}$.

Our next job is to escape these limitations. First we show that $v[\phi, \bar{K}, p]$ is a kind of estimate of the number of points in $\bar{K}$ which are mapped by ϕ onto p.

DEFINITION. Let ϕ be a continuous mapping of a set $S \subset R^n$ into R^n where R^n is oriented. Suppose p is a point of S and $\phi(p) = q$. Then p is a *q-point under ϕ*. Assume further that there is a neighborhood $N_\varepsilon(p)$ such that if for $x \in N_\varepsilon(p)$ it is true that $\phi(x) = q$, then $x = p$. Then p is an *isolated q-point under ϕ*.

REMARK. If all the q-points are isolated and S is bounded, the set of q-points is finite.

LEMMA. *Let x_1^n, x_2^n be two positively oriented n-simplexes in R^n such that p is an interior point of $\bar{x}_i^n$ $(i = 1, 2)$ and suppose $x_i^n \subset N_\varepsilon(p)$, the neighborhood in the definition of isolated q-point, for $i = 1, 2$. Then $v(\phi, \overline{b(x_i^n)}, q)$ is defined for $i = 1, 2$ and*

$$v(\phi, \overline{b(x_1^n)}, q) = v(\phi, \overline{b(x_2^n)}, q).$$

PROOF. Since $b(x_i^n)$ is a cycle and p is an isolated q-point in $N_\varepsilon(p)$ where $N_\varepsilon(p) \supset \overline{x_i^n}$ then $v(\phi, \overline{b(x_i^n)}, q)$ is defined for $i = 1, 2$. Using the point p as the center of projection, project from $\overline{b(x_1^n)}$ onto $\overline{b(x_2^n)}$ and denote the projection mapping by P. Then from the definition of order, it follows that:

$$v(\phi, \overline{b(x_2^n)}, q) = v(\phi P, \overline{b(x_1^n)}, q).$$

But

$$v(\phi P, \overline{b(x_1^n)}, q) = v(\phi, \overline{b(x_1^n)}, q)$$

by the Invariance under Homotopy Theorem (4.1) if we use as the homotopy ϕP_t on $b(x_1^n)$ where P_0 is the identity mapping and $P_t(y)$ is the point at distance $td(y, P(y))$ along the line segment $[y, P(y)]$ joining $y \in \overline{b(x_1^n)}$ and the projection $P(y)$ on $\overline{b(x_2^n)}$. (Thus $P_1(y) = P(y)$.) ∎

This lemma justifies the following definition.

(5.1) DEFINITION. If p is an isolated q-point and x^n is a positively oriented n-simplex such that p is an interior point of $\bar{x}^n$ and $\bar{x}^n \subset N_\varepsilon(p)$, then $v(\phi, \overline{b(x^n)}, q)$ is the *index of the q-point p*. We denote the index by $j(\phi, p, q)$.

DEFINITION. Let ϕ be a continuous mapping from $\bar{K}$ into R^n, oriented, where K is an n-dimensional complex with all its n-simplexes positively oriented in R^n. Let $q \in \phi(\bar{K})$. A q-point p is *regular* if p is in the interior of an n-simplex in K and p is an isolated q-point. (Note that the index of a regular q-point is defined.)

In the following, let K be an n-dimensional simplicial complex in an oriented R^n with its n-simplexes $x_1^n, \cdots, x_w^n$ positively oriented in R^n. We will deal with the chain $\sum_{i=1}^{w} x_i^n$ and its boundary. The notions of algebraic number of q-points and local degree at q (which we are about to define) can be defined more generally by including a wider class of chains on K, but in applications the definitions we will give are the only ones used.

DEFINITION. Let ϕ be a continuous mapping from $\bar{K}$ into R^n such that for a given point $q \in R^n$ all q-points of ϕ are regular. Let $p_1, \cdots, p_m$ be the set of q-points of ϕ and let $j^h = j(f, p_h, q)$ for $h = 1, \cdots, m$. The *algebraic number of q-points of ϕ in $\bar{K}$* is the sum

$$J = \sum_{h=1}^{m} j^h.$$

(5.2) THEOREM. *Let ϕ be a continuous mapping from $\bar{K}$ into R^n and suppose ϕ has only regular q-points for a given point $q \in R^n$. Then*

$$\sum_{h=1}^{m} j^h = v\left(\phi, \overline{b\left(\sum_{i=1}^{w} x_i^n\right)}, q\right).$$

PROOF. Let K_1, consisting of simplexes $y_1^n, \cdots, y_\mu^n$ and their sides, be a subdivision of K such that each q-point p_h is an interior point of an n-simplex of K_1 and such that each n-simplex of K_1 contains at most one point p_h. Let $y_1^n, \cdots, y_m^n$ be the n-simplexes of K_1 which contain the regular q-points $p_1, \cdots, p_m$.

LEMMA. $v(\phi, \overline{b(\sum_{i=1}^{w} x_i^n)}, q) = v(\phi, \overline{b(\sum_{i=1}^{\mu} y_i^n)}, q)$.

PROOF. Since all the simplexes in K and K_1 are positively oriented, then if $\{\sigma_j^{n-1}\}$ is the collection of $(n-1)$-simplexes which are contained in $\overline{b(\sum_{i=1}^{w} x_i^n)}$ and if $\{\tau_k^{n-1}\}$ is the collection of $(n-1)$-simplexes contained in $\overline{b(\sum_{i=1}^{\mu} y_i^n)}$, then

$$\bigcup_j \bar{\sigma}_j^{n-1} = \bigcup_k \bar{\tau}^{n-1}$$

because $\sigma_j^{n-1} \in \{\sigma_j^{n-1}\}$ if and only if each point in σ_j^{n-1} is a boundary point of $\bar{K}$. The proof then follows from Theorem (2.7). This completes the proof of the lemma.

From the definition of order and intersection number, it follows that

$$v\left(\phi, \overline{b\left(\sum_{i=1}^{\mu} y_i^n\right)}, q\right) = v\left(\phi, \overline{b\left(\sum_{h=1}^{m} y_h^n\right)}, q\right).$$

Again by the definition of order,

$$v\left(\phi, \overline{b\left(\sum_{h=1}^{m} y_h^n\right)}, q\right) = \sum_{h=1}^{m} v(\phi, \overline{b(y_h^n)}, q).$$

The expression on the right is exactly $\sum_{h=1}^{m} j^h$. ∎

(5.3) COROLLARY. *If ϕ_1, ϕ_2 are continuous mappings from $\bar{K}$ into R^n and the q-points of ϕ_1 and ϕ_2 are all regular and if $p \in b(\sum_{i=1}^{w} x_i^n)$ implies:*

$$|\phi_1(p) - \phi_2(p)| < |\phi_1(p) - q|$$

then ϕ_1, ϕ_2 have the same algebraic number of q-points in $\bar{K}$.

PROOF. Follows from Theorem (5.2) by application of the Poincaré-Bohl Theorem (4.2). ∎

(5.4) THEOREM. *If ϕ is a continuous mapping from $\bar{K}$ into R^n and q is an arbitrary point in R^n and $\varepsilon > 0$, then there is a continuous mapping ϕ_ε such that if $p \in \bar{K}$, then*

$$|\phi_\varepsilon(p) - \phi(p)| < \varepsilon$$

and such that the q-points of ϕ_ε are all regular.

PROOF. First assume that the set of q-points $p_1, \cdots, p_m$ of ϕ is a finite set. Then each q-point is isolated and there is a set of pairwise disjoint neighborhoods $N_{\delta_1}(p_1), \cdots, N_{\delta_m}(p_m)$. If a q-point p_i is on the boundary of an n-simplex of K, it is not difficult to construct a mapping ϕ_1 which is the same as ϕ except on N_i, such that for $p \in N_i$,

$$|\phi(p) - \phi_1(p)| < \varepsilon$$

and such that ϕ_1 has only a finite set of q-points in N_i and none of these q-points is on the boundary of an n-simplex of K. The mapping $\phi_1$would be the desired mapping except that there may be more than one q-point in an n-simplex. Now we show that there is a finite succession of simplicial subdivisions of K such that each n-simplex of the resulting complex contains at most one q-point. If $\bar{x}^n$ contains more than one q-point, let o be the center of gravity of $\bar{x}^n$. If $N_{\varepsilon_1}(o)$ is a neighborhood of o of radius ε_1, there is a point $o' \in N_{\varepsilon_1}(o)$ such that if o' is the point used in the construction described in the proof of Theorem (3.1), then each q-point in $\bar{x}^n$ is in the interior of one of the simplexes of the subdivision. Let δ be less than the minimum of $\delta_1, \delta_2, \cdots, \delta_m$. Continue to make subdivisions as described above until the diameters of the resulting simplexes are all less than δ. (The diameters can be made less than δ if we choose ε_1 sufficiently small. Cf. Theorem (3.3).) This completes the proof for the case in which the set of q-points is finite.

If the set of q-points is infinite, we need only show that there is a mapping ϕ_2 such that for $p \in \bar{K}$,

$$|\phi(p) - \phi_2(p)| < \frac{\varepsilon}{2}$$

and such that ϕ_2 has a finite set of q-points. Then the previous discussion can be applied to ϕ_2. Since $\bar{K}$ is a closed bounded set, the mapping ϕ is uniformly continuous on $\bar{K}$. So there is a number δ such that if

$$|p^{(1)} - p^{(2)}| < \delta,$$

then

$$|\phi(p^{(1)}) - \phi(p^{(2)})| < \frac{\varepsilon}{4}.$$

By Theorem (3.3), if δ is a given positive number, there is a simplicial subdivision of $\bar{K}$ such that the diameters of the simplexes in the subdivision are all less than δ. Now define function ϕ_L on the vertices V_i of the simplexes in the simplicial subdivision as:

$$\phi_L(V_i) = \phi(V_i)$$

unless the points $\phi(V_0), \phi(V_1), \cdots, \phi(V_n)$, where $V_0, V_1, \cdots, V_n$ are the vertices of an n-simplex of the simplicial subdivision of K, are linearly dependent. Then change each of the points $\phi_L(V_0), \phi_L(V_1), \cdots, \phi_L(V_n)$ so that they differ from the original $\phi_L(V_0), \cdots, \phi_L(V_n)$ by less than $\varepsilon/4$ and so that they are linearly independent. Then extend ϕ_L linearly over the simplexes of $\bar{K}$. The resulting mapping is the desired mapping ϕ_2. ∎

(5.5) THEOREM. *Let ϕ be a continuous mapping from $\bar{K}$ into an oriented R^n such that ϕ has no q-points on $\overline{b(\sum_{i=1}^{w} x_i^n)}$. Then there is an $\varepsilon > 0$ such that if ϕ_1, ϕ_2 are within ε of f for all points of $\bar{K}$ (i.e., if $p \in \bar{K}$, then $|\phi_i(p) - \phi(p)| < \varepsilon, i = 1, 2$) and if ϕ_1, ϕ_2 have only regular q-points, then the algebraic number of q-points of ϕ_1 is equal to the algebraic number of q-points of ϕ_2.*

PROOF. Since ϕ has no q-points on $\overline{b(\sum_{i=1}^{k} x_i^n)}$, then

$$d\left[\phi\left[\overline{b\left(\sum_{i=1}^{k} x_i^n\right)}\right], q\right] = \alpha > 0.$$

If ϕ_1, ϕ_2 are such that for all $p \in \overline{b(\sum_{i=1}^{k} x_i^n)}$,

(5.5-1) $$|\phi_i(p) - \phi(p)| < \tfrac{1}{3}\alpha \quad (i = 1, 2),$$

the theorem follows from Corollary (5.3). ∎

Theorems (5.4) and (5.5) justify the introduction of the following definition.

(5.6) DEFINITION. Let ϕ be a continuous mapping from $\bar{K}$ into R^n such that ϕ has no q-points on $\overline{b(\sum_{i=1}^{w}) x_i^n)}$. The *local degree of mapping ϕ at the point* q is the algebraic number of q-points of a continuous mapping ϕ_1 of $\bar{K}$ into R^n which has only regular q-points and which satisfies condition (5.5-1) above. The local degree is denoted by $d[\phi, \bar{K}, q]$.

(5.7) THEOREM. *If ϕ is a continuous mapping from $\bar{K}$ into R^n and if ϕ has no q-points on $\overline{b(\sum_{i=1}^{w} x_i^n)}$, then the local degree of ϕ at q is defined and is equal to $v(\phi, \overline{b(\sum_{i=1}^{w} x_i^n)}, q)$.*

PROOF. Follows from Definition (5.6), Theorem (5.2) and the fact that if ϕ_1 is sufficiently close to ϕ then by the Poincaré-Bohl Theorem (4.2),

$$v\left(\phi_1, \overline{b\left(\sum_{i=1}^{w} x_i^n\right)}, q\right) = v\left(\phi, \overline{b\left(\sum_{i=1}^{w} x_i^n\right)}, q\right).$$

6. **The local degree relative to the closure of a bounded open set.** So far we have defined the local degree of a continuous mapping ϕ where ϕ has domain $\bar{K}$. Now we extend the definition to include a continuous mapping ϕ with domain $\bar{D}$ where $\bar{D}$ is the closure of a bounded open set. This is done by "approximating" $\bar{D}$ by a complex K. Let ϕ be a continuous mapping of $\bar{D} \subset R^n$, an oriented Euclidean n-space, into R^n and let $q \in R^n$. Assume $q \notin \phi(\bar{D} - D)$.

(6.1) LEMMA. *There is an n-dimensional simplicial complex K such that:*

(6.1-1) $$\bar{K} \subset D;$$

(6.1-2) *if $\phi(p) = q$, then p is an interior point of $\bar{K}$.*

PROOF. First since D is open, the set $\bar{D} - D$ is closed. Also $\phi^{-1}(q)$ is a closed set because ϕ is continuous. The sets $\bar{D} - D$ and $\phi^{-1}(q)$ are also bounded. By hypothesis,

$$[\phi^{-1}(q)] \cap [\bar{D} - D] = \varnothing.$$

Hence since the sets $\phi^{-1}(q)$ and $\bar{D} - D$ are closed and bounded, then

$$d[\phi^{-1}(q), \bar{D} - D] = \delta > 0. \tag{6.1-3}$$

Since $\bar{D}$ is bounded, there is a rectangular parallelopiped R with sides parallel to the coordinate planes and such that $\bar{D} \subset R$. Subdivide R by hyperplanes parallel to the coordinate hyperplanes so that each rectangular parallelopiped in the subdivision has diameter less than $\delta/2$. Let $\{R_i\}$ be the collection of rectangular parallelopipeds each of which is contained in D. If $p \in \phi^{-1}(q)$, then p is in the interior of $\bigcup_i R_i$ because otherwise we have a contradiction to (6.1-3). Let K be the simplicial subdivision given by Theorem (3.1) of the cell complex consisting of the R_i and their sides. ▮

(6.2) LEMMA. *If K_1, K_2 are n-dimensional simplicial complexes satisfying condition* (6.1-1) *and* (6.1-2) *of Lemma* (6.1) *then*

$$d[\phi, \bar{K}_1, q] = d[\phi, \bar{K}_2, q].$$

PROOF. The intersection of two simplexes, one from K_1 and one from K_2, is a cell. The collection of all such cells is a cell complex $\mathscr{K}$ and $\bar{\mathscr{K}} = \bar{K}_1 \cap \bar{K}_2$. By Theorem (3.1) there is a simplicial subdivision K_3 of K, and K_3 is an n-dimensional simplicial complex. From the definition of local degree, it follows then that

$$d[\phi, \bar{K}_1, q] = d[\phi, \bar{K}_3, q]$$

and

$$d[\phi, \bar{K}_2, q] = d[\phi, \bar{K}_3, q]. ▮$$

Lemmas (6.1) and (6.2) justify the following definition.

(6.3) DEFINITION. Let D be a bounded open set in R^n, oriented, and ϕ a continuous mapping with domain $\bar{D}$ and range a subset of R^n. Let $q \in R^n$ and assume $q \notin \phi(\bar{D} - D)$. The *local degree of ϕ at q and relative to $\bar{D}$*, denoted by $d[\phi, \bar{D}, q]$, is equal to $d[\phi, \bar{K}, q]$ where K is an n-dimensional simplicial complex in R^n and the n-simplexes of K are oriented positively relative to R^n and K satisfies the conditions:

(1) $\bar{K} \subset D$;

(2) if $\phi(p) = q$, then p is an interior point of $\bar{K}$.

The following theorems are easy consequences of the corresponding theorems for $d[\phi, \bar{K}, q]$.

(6.4) THEOREM OF INVARIANCE UNDER HOMOTOPY. *Let ϕ_t be a continuous mapping from $\bar{D} \times [0, 1]$ and $q(t)$ the image in R^n of a continuous mapping* *from* $[0, 1]$ *such that for each* $t_0 \in [0, 1]$, *the point* $q(t_0) \notin \phi_{t_0}(\bar{D} - D)$. *Then for each* $t \in [0, 1]$, *the degree* $d(\phi_t, \bar{D}, q(t))$ *has the same value.*

(6.5) COROLLARY (POINCARÉ-BOHL THEOREM). *Let* ϕ_1, ϕ_2 *be continuous mappings from* $\bar{D}$ *into* R^n. *Suppose that for each* $p \in \bar{D} - D$, *the line segment* $\overline{\phi_1(p)\phi_2(p)}$ *does not intersect the point* $q \in R^n$. *Then*

$$d[\phi_1, \bar{D}, q] = d[\phi_2, \bar{D}, q].$$

(6.6) EXISTENCE THEOREM. *If* $d[\phi, \bar{D}, q] \neq 0$, *then there is a point* $p \in D$ *such that* $\phi(p) = q$.

Theorem (6.6) underlies all applications of local degree in analysis. Note, however, that Theorem (6.6) gives only a sufficient condition. That is, if $d[\phi, \bar{D}, q] = 0$, the equation:

$$\phi(x) = q$$

may have one or more solutions. We will give illustrations of this in Chapter II.

From Definitions (5.6) and (6.3), we have also the following simple observations:

(6.7) THEOREM. *If mappings* M_1 *and* M_2 *from* R^n *into* R^n *are defined by*:

$$M_1 : x \to f(x),$$
$$M_2 : x \to f(x) + p,$$

where f *is a continuous* n*-vector function and* p *is a constant* n*-vector and if* $d[M_2, \bar{D}, q]$ *is defined where* $\bar{D}$ *is the closure of a bounded open set in* R^n, *then* $d[M_1, \bar{D}, q - p]$ *is defined and*

$$d[M_1, \bar{D}, q - p] = d[M_2, \bar{D}, q].$$

(6.8) THEOREM. *If* $\bar{D}_1, \bar{D}_2$ *are the closures of bounded open sets* D_1 *and* D_2 *in* R^n *and if* $D_1 \cap D_2 = \varnothing$; *if* f *is a continuous mapping from* $\overline{D_1 \cup D_2}$ (*which is* $\bar{D}_1 \cup \bar{D}_2$) *into* R^n *and* p *is a point such that* $p \notin [f(\bar{D}_1 - D_1)] \cup [f(\bar{D}_2 - D_2)]$; *then*

$$d[f, \bar{D}_1, p] + d[f, \bar{D}_2, p] = d[f, \overline{D_1 \cup D_2}, p].$$

7. **The local degree as a lower bound for the number of solutions.** Our next objective is to show how the absolute value of the local degree of a mapping f yields a kind of lower bound for the number of points in the set $f^{-1}(q)$. The number $|d[f, \bar{D}, q]|$ is not, strictly speaking, a lower bound for the number of points in $f^{-1}(q)$. For example, if $\bar{D}$ is the solid circle of radius one and center at the origin 0 in the plane R^2 and if f is the mapping of R^2 into R^2 described in terms of complex variables by $f(z) = z^2$, then, as we will prove later, $d[f, \bar{D}, \bar{0}] = 2$. But $f^{-1}(\bar{0})$ consists of just one point, the origin $\bar{0}$. The theorem we will obtain says roughly that if f is differentiable

and the point q is changed, however slightly, to another point q_1, then $f^{-1}(q_1)$ consists of at least $|d[f, \bar{D}, q_1]|$ points. We will be able to make significant applications of this theorem to ordinary differential equations. The theorem is based on a theorem due to A. Sard [**1**] of which Lemma (7.3a) which follows is a simple version.

Definition. Let the mapping

$$f:(x_1, \cdots, x_n) \to (x_1', \cdots, x_n')$$

be described by the equations

$$x_i' = f_i(x_1, \cdots, x_n) \quad (i = 1, \cdots, n)$$

and suppose the partial derivatives $\partial f_i/\partial x_j$ $(i, j = 1, \cdots, n)$ exist and are continuous at each point of an open set $G \subset R^n$. Then f is *differentiable* on G.

Definition. The matrix

$$\left(\frac{\partial f_i}{\partial x_j}\right) = \begin{pmatrix} \dfrac{\partial f_1}{\partial x_1} & \cdots & \dfrac{\partial f_1}{\partial x_n} \\ \vdots & & \vdots \\ \dfrac{\partial f_n}{\partial x_1} & \cdots & \dfrac{\partial f_n}{\partial x_n} \end{pmatrix}$$

is called the *differential of* f. At each point $p_0 \in G$ the differential is denoted by $df]_{p_0}$.

Definition. The determinant $|\partial f_i/\partial x_j|$ is called the *Jacobian* of f. At each point $p_0 \in G$, the Jacobian is denoted by $Jf(p_0)$.

(7.1) Theorem. *Suppose f is differentiable on the open set $G \subset R^n$. Let p_0 be a fixed point in G. Then if $p \in G$,*

$$f(p) - f(p_0) = df\Big]_{p_0}(p - p_0) + R_{p_0}(p - p_0)$$

where

$$\lim_{|p - p_0| \to 0} \frac{R_{p_0}(p - p_0)}{|p - p_0|} = 0.$$

Also this limit is zero uniformly in p_0 on any closed bounded subset of G.

A proof of this theorem can be found in an advanced calculus text. For example, see Buck [**1**, pp. 184–186].

(7.2) Theorem. *Let f be a continuous mapping from $\bar{D}$ into R^n where D is an open set such that f is differentiable on D. Suppose $q \in R^n$. If $f(p_0) = q$ and $df]_{p_0}$ is a nonsingular matrix, then p_0 is an isolated q-point under f and the index of the q-point p_0 is $+1$ or -1 depending on whether $Jf(p_0)$ is positive or negative.*

PROOF. We first show that p_0 is an isolated q-point. Suppose there is a sequence of points $\{p_n\}$ such that $\lim_{n \to \infty} p_n = p_0$ and such that $f(p_n) = q$ for all n. Using Theorem (7.1) we have:

$$0 = f(p_n) - f(p_0) = df\Big]_{p_0} (p_n - p_0) + R_{p_0}(p_n - p_0).$$

Dividing the equation by $|p_n - p_0|$ we obtain:

$$(7.2\text{-}1) \qquad 0 = df\Big]_{p_0} \frac{(p_n - p_0)}{|p_n - p_0|} + \frac{R_{p_0}(p_n - p_0)}{|p_n - p_0|}.$$

Since $df]_{p_0}$ is nonsingular, there is a positive constant r such that for all n,

$$(7.2\text{-}2) \qquad \left| df\Big]_{p_0} \frac{(p_n - p_0)}{|p_n - p_0|} \right| \geqq r.$$

By Theorem (7.1), there is an $\varepsilon_0 > 0$ such that if $|p_n - p_0| < \varepsilon_0$, then

$$(7.2\text{-}3) \qquad \left| \frac{R_{p_0}(p_n - p_0)}{|p_n - p_0|} \right| < \frac{r}{2}.$$

For sufficiently large n, inequalities (7.2-2) and (7.2-3) contradict (7.2-1). Hence p_0 is an isolated q-point. Now let x^n be an n-simplex such that $\bar{x}^n$ is contained in D, let p_0 be an interior point of $\bar{x}^n$ and assume p_0 is the only q-point in $\bar{x}^n$. If the diameter of $\bar{x}^n$ is sufficiently small, then by (7.2-2) and (7.2-3) above and the Poincaré-Bohl Theorem (6.5),

$$v(f, \overline{b(x^n)}, q) = v\left(df\Big]_{p_0}, \overline{b(x^n)}, q\right).$$

From the definition of orientation in R^n and the definition of intersection number, it follows that $v(df]_{p_0}, \overline{b(x_n)}, 0)$ is $+1$ or -1 depending on whether $Jf(p_0)$ is positive or negative. ∎

Now we obtain the theorem which describes in what sense $|d[f, \bar{D}, p]|$ is a lower bound for the number of points in the set $f^{-1}(p)$.

(7.3) THEOREM. *Let f be continuous and differentiable on an open set $G \subset R^n$ which contains $\bar{D}$ the closure of a bounded open set and suppose the range of f is contained in a subset of R^n. Suppose $\mathscr{C}$ is a component of the open set $R^n - f(\bar{D} - D)$. Then there is a set $\mathscr{E}$ of n-measure zero such that $\mathscr{E} \subset \mathscr{C}$ and such that if $q_1 \in \mathscr{C} - \mathscr{E}$, then $f^{-1}(q_1)$ consists of a finite number m of points where $m \geqq |d[f, \bar{D}, q_1]|$ and*

$$m = d[f, \bar{D}, q_1] \pmod 2.$$

Before proving the theorem, we point out the following corollary which is the form in which the theorem will be used most frequently.

(7.4) COROLLARY. *If $q \in R^n - f(\bar{D} - D)$, then there exists a neighborhood $N_\varepsilon(q)$ and a set E of n-measure zero such that $E \subset N_\varepsilon(q)$ and such that if*

$q_1 \in N_\varepsilon(q) - E$, then $d[f, \bar{D}, q] = d[f, \bar{D}, q_1]$ and $f^{-1}(q_1)$ consists of a finite number m of points where $m \geqq |d[f, \bar{D}, q]|$ and $m = d[f, \bar{D}, q]$ (mod 2).

PROOF. The proof follows from Theorem (7.3), the Invariance under Homotopy Theorem (6.4), and the facts that the components of an open set in a locally connected space are open and a subset of a set of measure zero has measure zero.

For the proof of the theorem, we need some preliminaries:

DEFINITION. Let f be a differentiable mapping of an open set $G \subset R^n$ into R^n. A point $x \in G$ at which the Jacobian of f is zero is a *critical point* of f.

(7.3a) LEMMA. *Let f be a differentiable mapping of a bounded open set $G \subset R^n$ into R^n. Let D be an open set such that $\bar{D} \subset G$. Then the image under f of the set of critical points in D has n-measure zero.*

PROOF. Since G is bounded, there is a rectangular parallelopiped R with sides parallel to the coordinate planes such that $G \subset R$. Now divide R by planes parallel to the coordinate planes so that the diameters of the small rectangular parallelopipeds or cubes thus obtained are $< \frac{1}{2}d(\bar{D}, G^c)$. Then $\bar{D}$ is contained in the union of a finite number N of cubes r_i and each r_i is contained in G. We shall prove that the image of the set of critical points in $\bigcup_i r_i$ has n-measure zero. Let r_i be such a cube with side-length l. By Theorem (7.1), $\varepsilon > 0$ implies that there exists an integer m such that if $p_1, p_2 \in r_i$ and if $|p_1 - p_2| < l\sqrt{n}/m$, then:

$$\frac{|R_{p_2}(p_1 - p_2)|}{|p_1 - p_2|} < \varepsilon$$

or $|R_{p_2}(p_1 - p_2)| < \varepsilon|p_1 - p_2| < \varepsilon l\sqrt{n}/m$. Now divide r_i into m^n cubes of side l/m. Call these cubes r_{ij}. For each $p_1, p_2 \in r_{ij}$,

$$f(p_1) - f(p_2) = df\Big]_{p_2} (p_1 - p_2) + R_{p_2}(p_1 - p_2)$$

where $|R_{p_2}(p_1 - p_2) < \varepsilon l\sqrt{n}/m$. Take p_2 fixed in r_{ij}. Let $T(r_{ij})$ be the image of r_{ij} under the transformation

$$T: p_1 \to f(p_2) + df\Big]_{p_2} (p_1 - p_2),$$

and let $L = \max_{p \in r_{ij}} \sum_{q,s=1}^{n} |\partial f_q(p)/\partial x_s|$. The diameter of r_{ij} is $l\sqrt{n}/m$. So the diameter of $T(r_{ij})$ is less than or equal to $L(l\sqrt{n}/m)$. Now suppose r_{ij} contains a critical point p_1. Then $T(r_{ij})$ is contained in an $(n - 1)$-space and all the points of $f(r_{ij})$ stay within $\varepsilon l\sqrt{n}/m$ of the set $T(r_{ij})$. Thus the n-measure of $f(r_{ij})$ is less than or equal to:

$$\left\{L\left(\frac{l\sqrt{n}}{m} + \frac{2\varepsilon l\sqrt{n}}{m}\right)\right\}^{n-1} \left\{\frac{2\varepsilon l\sqrt{n}}{m}\right\}.$$

Therefore

$$\sum_j \{\text{meas}\,[f(r_{ij})]\} \leqq m^n \left\{L\left(\frac{l\sqrt{n}}{m} + \frac{2\varepsilon l\sqrt{n}}{m}\right)\right\}^{n-1} \left\{\frac{2\varepsilon l\sqrt{n}}{m}\right\}$$
$$= L^{n-1}(l\sqrt{n} + 2\varepsilon l\sqrt{n})^{n-1}(2\varepsilon l\sqrt{n}).$$

Quantities L, l, n are fixed. Hence if ε is sufficiently small, the right side can be made as small as desired. Hence the image under f of the set of critical points of f on D has n-measure zero. This completes the proof of Lemma (7.3a).

To prove Theorem (7.3), let E be the image under f of the set of critical points and let $\mathscr{E} = E \cap \mathscr{C}$. By the lemma, the set $\mathscr{E}$ has n-measure zero. Let $q \in \mathscr{C} - \mathscr{E}$. Then at each point $p \in f^{-1}(q)$, the Jacobian of f is nonzero. Hence by Theorem (7.2), each q-point is isolated and hence the set S of q-points is finite. Also the index of each point of S is $+1$ or -1. By Definitions (5.6) and (6.3) of local degree, the local degree of f at q is the sum of the indices of the points of S. The conclusion of the theorem then follows. ∎

8. **A product theorem for local degree.** To facilitate the computation of the local degree of certain mappings, we obtain a theorem about the degree of a product mapping.

(8.1) Product Theorem. *Let f be a mapping, differentiable on a bounded open set G such that $G \supset \bar{D}$ the closure of an open set D in R^n, into R^n. Let g be a mapping, differentiable on a bounded open set U such that $U \supset f(\bar{D})$, into R^n. Let $q \in R^n$. Suppose the set of q-points of g is a finite set $p_1, \cdots, p_m$ and suppose $p_i \notin f(\bar{D} - D)$ for $i = 1, \cdots, m$. Then*

$$d[gf, \bar{D}, q] = \sum_{i=1}^{m} \{d[f, \bar{D}, p_i]\}\{j(g, p_i, q)\}$$

where $j(g, p_i, q)$ is the index of q-point p_i (Definition (5.1)).

Proof. First, the hypotheses guarantee that $d[gf, \bar{D}, q]$ is defined and that $d[f, \bar{D}, p_i]$ and $j[g, p_i, q]$ are defined for $i = 1, \cdots, m$.

The case in which p_i is not a critical point of g for $i = 1, \cdots, m$ follows from the definition of local degree (Definitions (5.6) and (6.3)), i.e., we have at once:

$$d[gf, \bar{D}, q] = \sum_{i=1}^{m} \{d[f, \bar{D}, p_i]\}\{j(g, p_i, q)\}.$$

In the general case there is, by the Invariance under Homotopy Theorem (6.4), a neighborhood $N_{\varepsilon_i}(p_i)$ such that $d[f, \bar{D}, p]$ is constant for $p \in N_{\varepsilon_i}(p_i)$. Also $N_{\varepsilon_i}(p_i)$ may be chosen so that the only q-point of g in $N_{\varepsilon_i}(p_i)$ is p_i. Let $N_\delta(q)$ be a neighborhood of q such that $d[gf, \bar{D}, q']$ is constant for $q' \in N_\delta(q)$. By Corollary (7.4), there is a $q' \in N_\delta(q)$ such that $g^{-1}(q')$ contains no critical points but

$$g^{-1}(q') \subset \bigcup_i N_{\varepsilon_i}(p_i).$$

Thus $g^{-1}(q')$ is a finite number of points $q_{i1}, \cdots, q_{ia_i}$, $(i = 1, \cdots, m)$ and, for $b = 1, \cdots, a_i$:

$$j(g, q_{ib}, q') = +1 \text{ or } -1.$$

Finally

$$\sum_{b=1}^{a_i} j(g, q_{ib}, q') = j(g, p_i, q).$$

Then

$$\begin{aligned} d[gf, \bar{D}, q] &= d[gf, \bar{D}, q'] \\ &= \sum_{i,b} \{d[f, \bar{D}, q_{ib}]\}\{j(g, q_{ib}, q')\} \\ &= \sum_{i,b} \{d[f, \bar{D}, p_i]\}\{j(g, q_{ib}, q')\} \\ &= \sum_{i} \{d[f, \bar{D}, p_i]\}\{j(g, p_i, q)\}. \blacksquare \end{aligned}$$

9. **Computation of the local degree.** The problem of computing the local degree is, to a considerable extent, unsolved except in the plane. We describe a computational method for mappings in the plane and the computation for several special classes of mappings of R^n into R^n with $n > 2$.

(9.1) THE LOCAL DEGREE OF SOME MAPPINGS FROM AN INTERVAL INTO THE REAL LINE. First we point out that in the one-dimensional case, the local degree can be easily computed for polynomial mappings. Let M be the mapping

$$M : x \to ax^n$$

where a is a constant and n a positive integer. Then if I is a closed interval such that 0 is an interior point of I, it follows from Definition (6.3) of local degree and the Product Theorem (8.1) that

$$\begin{aligned} d[M, I, 0] &= 0 && \text{if } n \text{ is even,} \\ d[M, I, 0] &= 1 && \text{if } n \text{ is odd and } a \text{ is positive,} \\ d[M, I, 0] &= -1 && \text{if } n \text{ is odd and } a \text{ is negative.} \end{aligned}$$

Now let M_2 be defined by

$$M_2 : x \to a_n x^n + \cdots + a_1 x + a_0$$

and let I be an interval $[-x_0, x_0]$ with center 0. If $|x_0|$ is sufficiently large, then

$$|a_n x_0^n - (a_n x_0^n + \cdots + a_1 x_0 + a_0)| < |a_n x_0^n|.$$

Hence by the Poincaré-Bohl Theorem (6.5), if the interval I is sufficiently large,

$$d[M_2, I, 0] = d[M, I, 0].$$

Extensions of this result can be made using Theorems (6.7) and (6.8).

(9.2) THE LOCAL DEGREE OF MAPPINGS IN THE PLANE. First let $M:(x, y) \to (x', y')$ be a mapping of R^2 into R^2 defined by:

$$x' = P(x, y),$$
$$y' = Q(x, y)$$

where P and Q are polynomials homogeneous in x and y of degrees m and n respectively. Let $\bar{0}$ be the origin of R^2 and let:

$$S_1 = [(x, y) \,/\, x^2 + y^2 \leqq 1].$$

We assume that $d[M, S_1, \bar{0}]$ is defined, i.e., we assume that there is no point p on the boundary of S_1 such that

$$M(p) = \bar{0}.$$

This means we assume that P and Q have no common real linear factors. Because of this assumption, the only $\bar{0}$-point of M is $\bar{0}$ itself. Since P and Q are homogeneous, it follows that for every solid circle S with center $\bar{0}$, the degree $d[M, S, \bar{0}]$ is defined and $d[M, S, \bar{0}]$ has the same value for all S.

Now write P and Q in factored form, i.e.,

$$x' = K_1 \prod_{i=1}^{m} (y - a_i x)^{p_i},$$
$$y' = K_2 \prod_{j=1}^{n} (y - b_j x)^{q_i}$$

where K_1, K_2 are constants. (We include the possibility that some $a_i = \infty$ or some $b_j = \infty$, equivalently, that factor $y - a_i x$ is equal to x or $y - b_j x$ is equal to x.) Let L be the linear mapping

$$L:(x'', y'') \to (x', y')$$

defined by

$$x' = K_1 x'',$$
$$y' = K_2 y''.$$

From the definition of local degree the local degree $d[L, S, \bar{0}]$ is $+1$ or -1 according as the sign of $K_1 K_2$ is positive or negative. Let M_1 be the mapping

$$M_1:(x, y) \to (x'', y'')$$

defined by:

$$x'' = \prod_{i=1}^{m} (y - a_i x)^{p_i},$$
$$y'' = \prod_{j=1}^{n} (y - b_j x)^{q_j}.$$

Since $d[M, S, \bar{0}]$ is defined then $d[M_1, S, \bar{0}]$ is defined and since $M = LM_1$, then by Theorem (8.1),

$$d[M_1, S, \bar{0}] = \pm d[M, S, \bar{0}].$$

Thus our problem is to compute $d[M_1, S, \bar{0}]$. We investigate the image under M_1 of the boundary of S by studying the changes of sign of x'' and y'' as (x, y) varies over the boundary of S. First we show that the following factors either do not change the signs of x'' or y'' or do not change them significantly, i.e., we will show that the value of $d[M_1, S, \bar{0}]$ is not affected if these factors are omitted:

(1) Pairs of factors $(y - a_{i_1}x)$, $(y - a_{i_2}x)$ in which a_{i_1} and a_{i_2} are complex conjugates. (Then product $(y - a_{i_1}x)(y - a_{i_2}x)$ is positive for all x and y.)

(2) Factors $(y - a_{i_1}x)^{p_i}$ where a_{i_1} is real and p_i is even.

(3) Pairs of factors $(y - a_{i_1}x)$, $(y - a_{i_2}x)$ such that $a_{i_1} < a_{i_2}$ and there is no b_j or a_{i_3} such that

$$a_{i_1} < b_j < a_{i_2}$$

or

$$a_{i_1} < a_{i_3} < a_{i_2}.$$

(4) Pairs of factors $(y - a_1x)$, $(y - a_mx)$ where a_1 and a_m are the smallest and largest of all the numbers $a_1, \cdots, a_m, b_1, \cdots, b_n$.

(5) The factors listed above with the a_i's replaced by b_j's.

It is fairly reasonable, intuitively, that we may disregard the factors listed above; nevertheless we give formal proofs which, incidentally, illustrate the use of the Invariance under Homotopy Theorem (6.4). We describe the homotopies that are used. It is easy to show that these homotopies satisfy the hypotheses of Theorem (6.4).

(1) Use the homotopy

$$M_t : (x, y) \to (x_1, y_1)$$

defined by

$$x_1 = \left[\prod_{i=1,\, i \neq i_1,\, i \neq i_2}^{m} (y - a_ix)^{p_i}\right] [t(y - a_{i_1}x)(y - a_{i_2}x) + (1 - t)],$$

$$y_1 = \prod_{j=1}^{n} (y - b_jx)^{q_j}.$$

(2) Use the homotopy M_t defined by:

$$x_1 = \left[\prod_{i=1,\, i \neq i_1}^{m} (y - a_ix)^{p_i}\right] [t(y - a_{i_1}x)^{p_{i_1}} + (1 - t)],$$

$$y_1 = \prod_{j=1}^{n} (y - b_jx)^{q_i}.$$

(3) Use the homotopy M_t defined by:

$$x_1 = \left[\prod_{i=1,\, i\neq i_1,\, i\neq i_2}^{m} (y - a_i x)^{p_i}\right] [y - a_{i_1}x][y - \{(1-t)a_{i_2} + t a_{i_1}\}x],$$

$$y_1 = \prod_{j=1}^{n} (y - b_j x)^{q_j}.$$

This reduces the problem to case (2) above.

(4) Use the homotopy M_t defined by:

$$x_1 = \left[\prod_{i=2}^{m-1} (y - a_i x)^{p_i}\right] [F(x, y, t)]\,[(1-t)y - a_m x],$$

$$y_1 = \prod_{j=1}^{n} (y - b_j x)^{q_j}$$

where, if $a_1 < 0$, $F(x, y, t) = (1-t)y - a_1 x$. When $t = 1$, we have again (2) above. If $a_1 = 0$, let $F(x, y, t) = y - t(-\varepsilon)x$ where $\varepsilon > 0$, and we have again the case $a_1 < 0$. If $a_1 > 0$ let $F(x, y, t) = y - (1-t)a_1 x$ and we have the case $a_1 = 0$.

If all the factors in $\prod_{i=1}^{m} (y - a_i x)^{p_j}$ or in $\prod_{j=1}^{n} (y - b_j x)^{q_j}$ are included in the above four classifications, then the local degree is zero because the mapping is homotopic to a mapping $M:(x, y) \rightarrow (x_2, y_2)$ where x_2 or y_2 is a constant.

To complete the study we need only consider a mapping M_1 such that

$$a_1 < b_1 < a_2 < \cdots < a_m < b_m$$

or

$$b_1 < a_1 < b_2 < \cdots < b_m < a_m$$

and with each factor $(y - a_i x)$ and $(y - b_j x)$ having exponent one. For definiteness, take the first case. Start on the boundary of S_1 at $(0, -1)$ and proceed in a counterclockwise direction. All the points on the arc until the line $y = a_1 x$ is encountered are mapped into the same quadrant. The points on the arc between $y = a_1 x$ and $y = b_1 x$ are mapped into the quadrant with the y-coordinate having the same sign and the x-coordinate having the opposite sign. Proceeding in this manner around the boundary of S_1 and observing the changes in sign and using Theorem (2.7), we see that if

$$a_1 < b_1 < a_2 < \cdots < a_m < b_m,$$

the local degree is m and if

$$b_1 < a_1 < \cdots < b_m < a_m,$$

the local degree is $-m$.

Now we compute $d[M, S, \bar{0}]$ for a mapping $M:(x, y) \to (x', y')$ defined by:

$$x' = \sum_{i=0}^{m} P_i(x, y),$$

$$y' = \sum_{j=0}^{n} Q_j(x, y)$$

where $P_i(x, y)$ and $Q_j(x, y)$ are homogeneous in x and y of degrees i and j respectively. Let $M_1:(x, y) \to (x'', y'')$ be defined by:

$$x'' = P_m(x, y),$$

$$y'' = Q_n(x, y).$$

We assume that P_m and Q_n have no common real linear factors so that $d[M_1, S, \bar{0}]$ is defined for all circles S with center $\bar{0}$ and can be computed by the preceding method. Since P_m and Q_n have no common real linear factors, then if

$$C_1 = [(x, y) | x^2 + y^2 = 1]$$

there is an $\varepsilon_1 > 0$ and collection of arcs $\sigma_1, \sigma_2, \cdots, \sigma_s$ on C_1 subtended by angles $\alpha_1, \cdots, \alpha_s$ such that $\bigcup_{i=1}^{s} \sigma_i = C_1$ and $\sigma_1, \cdots, \sigma_s$ are pairwise disjoint or have at most common endpoints and such that on each σ_i, either

$$|P_m(x, y)| > \varepsilon_1$$

or

$$|Q_n(x, y)| > \varepsilon_1.$$

Suppose $|P_m(x, y)| > \varepsilon_1$ on arc σ_k. Writing $\sum_{i=0}^{m} P_i(x, y)$ in polar coordinates:

$$\sum_{i=0}^{m} P_i(x, y) = \sum_{i=0}^{m} r^i P_i(\cos\theta, \sin\theta)$$

where $P_i(\cos\theta, \sin\theta)$ is homogeneous of degree i in $\cos\theta$ and $\sin\theta$ and by assumption $|P_m(\cos\theta, \sin\theta)| > \varepsilon_1$ for θ in the arc σ_κ (more precisely if the terminal side of angle θ intersects arc σ_κ), we see that if S_R is a circle with center at the origin and sufficiently large radius R, then on the arc $\sigma_{R\kappa}$ on S_R which is subtended by angle α_κ (see Figure 8),

$$\left| \sum_{i=0}^{m-1} P_i(x, y) \right| < \tfrac{1}{2} |P_m(x, y)|.$$

Then if $(x, y) \in \sigma_{R\kappa}$, the line joining $(P_m(x, y), Q_n(x, y))$ and $(\sum_{i=0}^{m} P_i(x, y), \sum_{j=0}^{n} Q_i(x, y))$ does not pass through the origin. (See Figure 8.) On each arc $\sigma_{R1}, \cdots, \sigma_{Rs}$, such an argument can be made for either $P_m(x, y)$ or $Q_n(x, y)$. Hence applying the Poincaré-Bohl Theorem (6.5), we may conclude: if circle S_R has a sufficiently large radius, then

$$d[M, S_R, \bar{0}] = d[M_1, S_R, \bar{0}].$$

Now let mapping $M:(x, y) \to (x', y')$ be defined by:

$$x' = P_m(x, y) + R_P(x, y),$$
$$y' = Q_n(x, y) + R_Q(x, y)$$

where $R_P(x, y)$ is a continuous function of (x, y) and is of order $m + 1$, i.e., $\lim_{r\to 0} R_P(x, y)/r^m = 0$ where $r^2 = x^2 + y^2$ and similarly R_Q is of order $n + 1$. As before, let mapping $M_1:(x, y) \to (x'', y'')$ be defined by:

$$x'' = P_m(x, y),$$
$$y'' = Q_n(x, y)$$

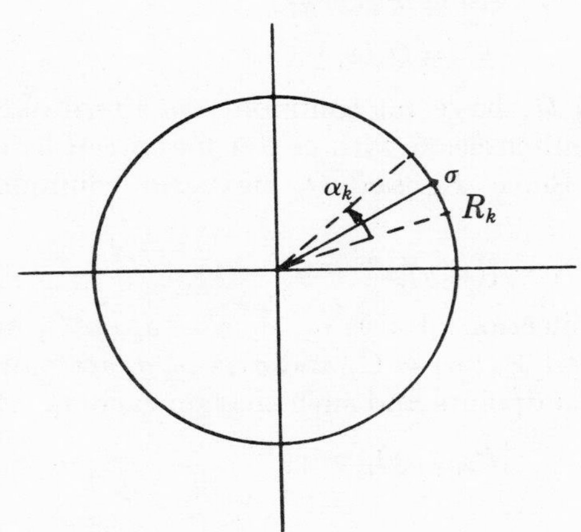

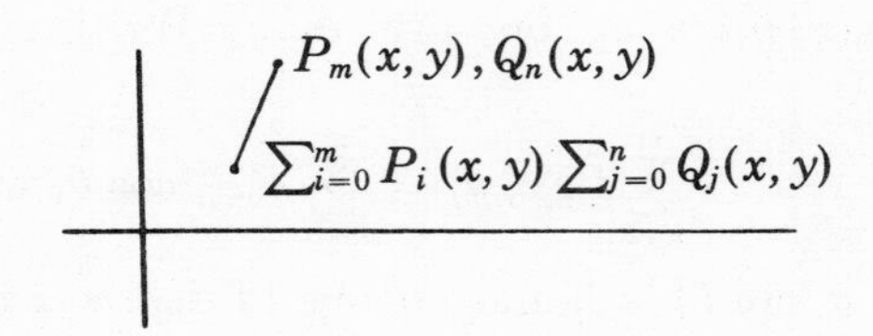

FIGURE 8

and assume P_m and Q_n have no common real linear factors. We show that if S_R is a circle of sufficiently small radius and center at the origin $\bar{0}$, then

$$d[M, S, \bar{0}] = d[M_1, S, \bar{0}].$$

Let $\sigma_1, \cdots, \sigma_s$ be the same arcs as before. Using polar coordinates, we write:

$$P_m(x, y) + R_P(x, y) = r^m \left[P_m(\cos\theta, \sin\theta) + \frac{R_p(x, y)}{r^m} \right].$$

Suppose $|P_m(x, y)| > \varepsilon_1$ on arc σ_K. Then for θ in arc σ_K,

$$|P_m(\cos\theta, \sin\theta)| > \varepsilon_1.$$

Hence if r is sufficiently small,

$$|R_P(x, y)| = \left| r^m \frac{R_P(x, y)}{r^m} \right| < \frac{r^m}{2} |P_m(\cos \theta, \sin \theta)|.$$

Thus if (x, y) is sufficiently close to $\bar{0}$ and (x, y) is in the angle α_κ which subtends arc σ_κ, the line joining $(P_m(x, y), Q_n(x, y))$ and

$$(P_m(x, y) + R_P(x, y), Q_n(x, y) + R_Q(x, y))$$

does not pass through the origin. Applying the Poincaré-Bohl Theorem (6.5), we conclude: if the radius of circle S_R is sufficiently small, then

$$d[M, S, \bar{0}] = d[M_1, S, \bar{0}].$$

(9.3) SOME MAPPINGS FROM R^{2n} INTO R^{2n}. So far we have computed the local degree by studying the behavior of the mapping on the boundary of the set, i.e., by studying the order of the point $\bar{0}$ relative to the image of the boundary. Now we compute some local degrees by regarding the degree as the algebraic number of q-points. We consider mappings from R^{2n} into R^{2n} which can be described in terms of functions of n complex variables: for example, the mapping $M : (x, y) \to (x', y')$ defined by

$$x' = x^2 - y^2,$$
$$y' = 2xy$$

can be expressed as $z' = z^2$ where $z = x + iy$, $z' = x' + iy'$.

First let M be the mapping of R^2 into R^2 described by:

$$z' = z^k$$

where k is a positive integer. Let S be a circle of center $\bar{0}$ (the origin) and arbitrary fixed radius r. To compute $d[M, S, \bar{0}]$, let $N_\varepsilon(\bar{0})$ be a neighborhood of $\bar{0}$ such that if $p \in N_\varepsilon(\bar{0})$, then $d[M, S, p]$ is defined and $d[M, S, p] = d[M, S, \bar{0}]$. (Theorem (6.4) guarantees the existence of neighborhood $N_\varepsilon(\bar{0})$.) For each point $p \neq \bar{0}$, the equation

$$z^k = p \tag{9.3-1}$$

has exactly k distinct solutions. If we write

$$z^k = u(x, y) + iv(x, y)$$

then the Jacobian of M at each of the solutions is

$$J = \begin{vmatrix} u_x & u_y \\ v_x & v_y \end{vmatrix}.$$

At each solution of (9.3-1), not all the derivatives u_x, u_y, v_x, v_y are zero. Thus by the Cauchy-Riemann equations, Jacobian J is positive. So by the Definition (6.3) of local degree and by Theorem (7.2), the local degree is k.

Now let M_j be the mapping of R^{2n} into R^{2n} described by:

$$\begin{aligned} z_1' &= z_1 \\ &\cdots \\ z_{j-1}' &= z_{j-1} \\ z_j' &= z_j^{k_j} \\ z_{j+1}' &= z_{j+1} \\ &\cdots \\ z_n' &= z_n. \end{aligned}$$

First $d[M_j, S, \overline{0}]$ is defined for all spheres S with center $\overline{0}$. Let $N_\varepsilon(\overline{0})$ be a neighborhood of $\overline{0}$ such that if $p \in N_\varepsilon(\overline{0})$, then $d[M_j, S, p] = d[M_j, S, \overline{0}]$. If $a \neq 0$, there are exactly k_j distinct solutions of

$$\begin{aligned} z_1 &= 0 \\ &\cdots \\ z_{j-1} &= 0 \\ z_j^{k_j} &= a \\ z_{j+1} &= 0 \\ &\cdots \\ z_n &= 0 \end{aligned}$$

and the Jacobian is positive at each solution. Hence the degree is k_j.

Now let the mapping, $\mathscr{M}$ of R^{2n} into R^{2n} be described by:

$$\begin{aligned} z_1' &= z_1^{k_1} \\ z_2' &= z_2^{k_2} \\ &\cdots \\ z_n' &= z_n^{k_n}. \end{aligned}$$

Then $\mathscr{M}$ is the product mapping $M_1 M_2 \cdots M_k$. Hence by Theorem (8.1): for arbitrary sphere S in R^{2n} with center $\overline{0}$,

$$d[\mathscr{M}, S, \overline{0}] = \prod_{j=1}^{n} k_j.$$

Finally let $P_{k_j}(z_1, \cdots, z_n)$ be a polynomial, homogeneous in $z_1, \cdots, z_n$, of degree k_j and let the mapping $\mathscr{M}_1$ from R^{2n} into R^{2n} be described by:

$$\begin{aligned} z_1' &= P_{k_1}(z_1, \cdots, z_n) \\ &\cdots \\ z_n' &= P_{k_n}(z_1, \cdots, z_n). \end{aligned}$$

Assume that $d[\mathscr{M}_1, S, \bar{0}]$ is defined for all spheres S with center $\bar{0}$. We will show that

$$d[\mathscr{M}_1, S, \bar{0}] = \prod_{i=1}^{n} k_i.$$

LEMMA. *Let $f(z_1, \cdots, z_q)$ be a polynomial with complex coefficients in the complex variables $z_1, \cdots, z_q$. Let*

$$V_f = [(z_1, \cdots, z_q) \in R^q / f(z_1, \cdots, z_q) = 0]$$

where R^q is a complex Euclidean q-space. Then if $p_1, p_2 \in R^q - V_f$, there is a continuous path in $R^q - V_f$ which has p_1 and p_2 as its endpoints.

PROOF. Define the polynomial in a complex variable λ:

$$\phi(\lambda) = f[(1-\lambda)p_1 + \lambda p_2].$$

Since $\phi(0) = f(p_1) \neq 0$ and $\phi(1) = f(p_2) \neq 0$, polynomial ϕ is not identically zero and hence has only a finite number of zeros. Since these do not include 0 or 1, there is in the λ-plane a continuous path λ_t $(0 \leqq t \leqq 1)$ such that $\lambda_0 = 0$, $\lambda_1 = 1$ and such that $\phi(\lambda_t) \neq 0$ for $0 \leqq t \leqq 1$. Hence $(1 - \lambda_t)p_1 + \lambda_t p_2$ is a path in $R^q - V_f$ of the desired type. ∎

The coefficients $a_1, a_2, \cdots, a_q$ of $P_{k_1}, \cdots, P_{k_n}$ may be regarded (if given some specified ordering) as a point in R^q. Thus each mapping defined by polynomials $P_{k_1}, \cdots, P_{k_n}$ of degrees $k_1, \cdots, k_n$ can be regarded as a point in R^q. Then $R(\mathscr{M}_1)$, the resultant of polynomials $P_{k_1}, \cdots, P_{k_n}$, is a polynomial in $a_1, a_2, \cdots, a_q$ with integer coefficients. The condition that $R(\mathscr{M}_1) \neq 0$ is equivalent to the condition that $d[\mathscr{M}_1, S, \bar{0}]$ be defined for all S (see van der Waerden [**1**, Vol. 2, p. 15]). Let L_0 and L_1 be two mappings defined by homogeneous polynomials of degrees $k_1, \cdots, k_n$ such that $R(L_0) \neq 0$ and $R(L_1) \neq 0$. By the lemma above, there is a continuous family of mappings L_t each defined by homogeneous polynomials of degree $k_1, \cdots, k_n$, and such that $R(L_t) \neq 0$ for $0 \leqq t \leqq 1$. Hence by the Invariance under Homotopy Theorem (6.4),

$$d[L_0, S, \bar{0}] = d[L_1, S, \bar{0}]$$

for all spheres S with center $\bar{0}$. But we have already shown that for a particular mapping of this type (the mapping $\mathscr{M}$) the local degree is $\prod_{j=1}^{n} k_j$. Therefore each mapping $\mathscr{M}_1$ such that $R(\mathscr{M}_1) \neq 0$ is such that

$$d[\mathscr{M}_1, S, \bar{0}] = \prod_{j=1}^{n} k_j$$

for all spheres S with center $\bar{0}$.

By using the same kind of arguments as in subsection (9.2) we have also:

if the resultant of polynomials $P_{k_1}, \cdots, P_{k_n}$ is nonzero and if $\mathscr{M}_2$ is the mapping of R^{2n} defined by:

$$z_1' = \sum_{i_1=0}^{k_1} P_{i_1}(z_1, \cdots, z_n)$$

$$\cdots$$

$$z_n' = \sum_{i_n=0}^{k_n} P_{i_n}(z_1, \cdots, z_n)$$

where $P_{i_j}(z_1, \cdots, z_n)$ is a polynomial homogeneous in $z_1, \cdots, z_n$ of degree i_j ($j = 1, \cdots, n$), then if S is a sphere with center $\bar{0}$ and S has sufficiently large radius, then

$$d[\mathscr{M}_2, S, \bar{0}] = d[\mathscr{M}_1, S, \bar{0}]$$

where $\mathscr{M}_1$ is the mapping defined by $P_{k_1}, \cdots, P_{k_n}$.

If $\mathscr{M}_3$ is the mapping of R^{2n} into R^{2n} defined by:

$$z_1' = P_{k_1}(z_1, \cdots, z_n) + P_{k_1+1}(z_1, \cdots, z_n)$$

$$\cdots$$

$$z_n' = P_{k_n}(z_1, \cdots, z_n) + P_{k_n+1}(z_1, \cdots, z_n)$$

where $P_{k_j}(z_1, \cdots, z_n)$ denotes the same kind of polynomial as before and P_{k_j+1} is a continuous function of $(z_1, \cdots, z_n)$ and is of order $k_j + 1$, i.e.,

$$\lim_{r \to 0} \frac{|P_{k_j+1}(z_1, \cdots, z_n)|}{r^{k_j}} = 0,$$

where $r^2 = |z_1|^2 + \cdots + |z_n|^2$, and if $R(\mathscr{M}_1) \neq 0$ then by the same kind of argument as in subsection (9.2), if S is a sphere with center $\bar{0}$ and sufficiently small radius, then

$$d[\mathscr{M}_3, S, \bar{0}] = d[\mathscr{M}_1, S, \bar{0}].$$

(9.3-2) LEMMA. *Suppose $\mathscr{M} : R^{2n} \to R^{2n}$ is differentiable and can be described by mapping $\mathfrak{M} : \mathscr{R}^n \to \mathscr{R}^n$ of complex Euclidean n-space $\mathscr{R}^n$ into itself where each component of $\mathfrak{M}$ is differentiable with respect to $z_1, \cdots, z_n$. Then the Jacobian of $\mathscr{M}$ is always non-negative.*

PROOF. Let $\mathscr{J}$ be the Jacobian of $\mathfrak{M}$, i.e., if $\mathfrak{M}$ is described by:

$$Z_1' = f_1(z_1, \cdots, z_n),$$

$$\cdots$$

$$Z_n' = f_n(z_1, \cdots, z_n),$$

then

$$J = \begin{vmatrix} \dfrac{\partial f_1}{\partial z_1} & \cdots & \dfrac{\partial f_1}{\partial z_n} \\ \vdots & & \vdots \\ \dfrac{\partial f_n}{\partial z_1} & \cdots & \dfrac{\partial f_n}{\partial z_n} \end{vmatrix}.$$

Let the corresponding matrix be put in rational canonical form, i.e., a basis in $\mathscr{R}^n$ is selected so that the corresponding matrix is:

$$\begin{pmatrix} \lambda_1 & 1 & & & & \\ & \lambda_1 & 1 & & & \\ & & \lambda_1 & 1 & & \\ & & & \ddots & & \\ & & & & \lambda_n & 1 \\ & & & & & \lambda_n \end{pmatrix}$$

where the unwritten terms are zero. If $\lambda_j = c_j + id_j$, then in the corresponding basis in R^{2n}, the matrix is

$$\begin{pmatrix} c_1 & -d_1 & 1 & 0 & & & & \\ d_1 & c_1 & 0 & 1 & & & & \\ & & c_1 & -d_1 & & & & \\ & & d_1 & c_1 & & & 1 & 0 \\ & & & & \ddots & & & \\ & & & & & & 0 & 1 \\ & & & & & & c_n & -d_n \\ & & & & & & d_n & c_n \end{pmatrix}.$$

The determinant J of this matrix is

$$J = (|\lambda_1|^2)^{m_1} \cdots (|\lambda_n|^2)^{m_n}$$

where m_j is the multiplicity of λ_j. ∎

Now consider again the mapping $\mathscr{M}_3 : R^{2n} \to R^{2n}$ described by

$$Z_1' = P_{k_1}(z_1, \cdots, z_n) + P_{k_1+1}(z_1, \cdots, z_n)$$

$$\cdots$$

$$Z_n' = P_{k_n}(z_1, \cdots, z_n) + P_{k_n+1}(z_1, \cdots, z_n)$$

and suppose $d[\mathscr{M}_3, S, \bar{0}]$ is defined where S is a fixed sphere with center at the origin. We will prove:

$$d[\mathscr{M}_3, S, \bar{0}] \geqq \prod_{i=1}^{n} k_i.$$

Suppose the resultant of $P_{k_1}, \cdots, P_{k_n}$ is nonzero. As shown before, if S_ε is a sufficiently small sphere with center at $\bar{0}$ then $d[\mathscr{M}_3, S_\varepsilon, \bar{0}] = \prod_{i=1}^n k_i$. Take S_ε such that $S_\varepsilon \subset S$. Now by Corollary (7.4) there exists a point p in a neighborhood of $\bar{0}$ such that

$$d[\mathscr{M}_3, S_\varepsilon, p] = d[\mathscr{M}_3, S_\varepsilon, \bar{0}]$$

and

$$d[\mathscr{M}_3, S, p] = d[\mathscr{M}_3, S, \bar{0}]$$

and such that p is not the image of a critical point of $\mathscr{M}_3$. From Lemma (9.3-2), it follows that if $q \in \mathscr{M}_3^{-1}(p)$, then $J(q) > 0$. Thus $d[\mathscr{M}_3, S, p]$ is equal to the number of points in $\mathscr{M}_3^{-1}(p) \cap S$ and $d[\mathscr{M}_3, S_\varepsilon, p]$ is equal to the number of points in $\mathscr{M}_3^{-1}(p) \cap S_\varepsilon$. Since $S_\varepsilon \subset S$, then

$$d[\mathscr{M}_3, S_\varepsilon, p] \leqq d[\mathscr{M}_3, S, p].$$

Summarizing these statements, we have

$$\prod_{i=1}^n k_i = d[\mathscr{M}_3, S_\varepsilon, \bar{0}] = d[\mathscr{M}_3, S_\varepsilon, p] \leqq d[\mathscr{M}_3, S, p] = d[\mathscr{M}_3, S, \bar{0}].$$

If the resultant of $P_{k_1}, \cdots, P_{k_n}$ is zero, then by changing the coefficients of $P_{k_1}, \cdots, P_{k_n}$ slightly, we obtain polynomials $P_{k_1}^{(1)}, \cdots, P_{k_n}^{(1)}$ the resultant of which is nonzero. If the coefficients of $P_{k_1}^{(1)}, \cdots, P_{k_n}^{(1)}$ are sufficiently close to those of $P_{k_1}, \cdots, P_{k_n}$ and the mapping $\mathscr{M}_4$ is described by

$$Z_1' = P_{k_1}^{(1)}(z_1, \cdots, z_n) + P_{k_1+1}(z_1, \cdots, z_n)$$

$$\cdots$$

$$Z_n' = P_{k_n}^{(1)}(z_1, \cdots, z_n) + P_{k_n+1}(z_1, \cdots, z_n),$$

then $d[\mathscr{M}_3, S, \bar{0}] = d[\mathscr{M}_4, S, \bar{0}]$ by the Poincaré-Bohl Theorem (6.5) if the radius of S is sufficiently small. Now apply the argument of the preceding paragraph to $d[\mathscr{M}_4, S, \bar{0}]$.

(9.4) Comparison of mappings from R^n into R^n and the corresponding complex mappings. Now let M be a mapping of R^n into R^n such that M can be extended to a mapping $\mathfrak{M}$ of the complex Euclidean n-space $\mathscr{R}^n$ into $\mathscr{R}^n$ which satisfies the hypotheses of Lemma (9.3-2). That is, suppose M is described by:

$$x_1' = f_1(x_1, \cdots, x_n)$$

$$\cdots$$

$$x_n' = f_n(x_1, \cdots, x_n)$$

where $f_1, \cdots, f_n$ are real-valued functions and assume that $f_1, \cdots, f_n$ can be extended to complex-valued functions, call these also $f_1, \cdots, f_n$, of the complex valuables $z_j = x_j + iy_j$ $(j = 1, \cdots, n)$. Then $\mathfrak{M}$ is described by:

$$z_1' = f_1(z_1, \cdots, z_n)$$
$$\cdots$$
$$z_n' = f_n(z_1, \cdots, z_n).$$

(If $f_1(x_1, \cdots, x_n), \cdots, f_n(x_1, \cdots, x_n)$ are polynomials or analytic functions in $x_1, \cdots, x_n$ then M satisfies this condition.) Let

$$s = [(x_1, \cdots, x_n)/x_1^2 + \cdots + x_n^2 \leqq 1],$$
$$S = [(Z_1, \cdots, Z_n) \,/\, |Z_1|^2 + \cdots + |Z_n|^2 \leqq 1],$$

and let $M^{(2)}$ be the mapping of R^{2n} into R^{2n} into itself described by $\mathfrak{M}$ and let $S^{(2)}$ be the set:

$$[(x_1, y_1, \cdots, x_n, y_n) \,/\, (x_1 + iy_1, \cdots, x_n + iy_n) \in S]$$

i.e., $S^{(2)}$ is the solid unit sphere with center $\bar{0}$ in R^{2n}. By Corollary (7.4) there is a point p in a neighborhood of $\bar{0}$ in R^n such that p is not the image of a critical point under M or $M^{(2)}$ and such that

$$d[M, s, \bar{0}] = d[M, s, p]$$

and

$$d[M^{(2)}, S^{(2)}, \bar{0}] = d[M^{(2)}, S^{(2)}, p].$$

Regarding $d[M, s, p]$ as the algebraic number of p-points, we may say: $|d[M, s, p]|$ is less than or equal to the number of points in $[M^{-1}(p)] \cap s$. But from Lemma (9.3-2) it follows that the number of points in $[M^{-1}(p)] \cap s$ is less than or equal to $d[M^{(2)}, S^{(2)}, p]$ which equals $d[M^{(2)}, S^{(2)}, \bar{0}]$. (Here p and $\bar{0}$ are regarded as points in R^{2n}.) Hence

$$|d[M, s, p]| \leqq d[M^{(2)}, S^{(2)}, \bar{0}]$$

and also the number of solutions of

$$f_1(x_1, \cdots, x_n) = p_1$$
$$\cdots$$
$$f_n(x_1, \cdots, x_n) = p_n$$

where $p_1, \cdots, p_n$ are the coordinates of the point p, is less than or equal to $d[M^{(2)}, S^{(2)}, \bar{0}]$.

Finally, we show that

$$d[M, s, \bar{0}] = d[M^{(2)}, S^{(2)}, \bar{0}] \pmod 2.$$

Let p be the point described before. Let r be the number of points in $M^{-1}(p) \cap s$. Then from the definition of local degree as the algebraic number of p-points, it follows that

$$d[M, s, \bar{0}] = d[M, s, p] = r \pmod 2.$$

To complete the proof, we show that

$$r = d[M^{(2)}, S^{(2)}, \bar{0}] \pmod 2.$$

Let E be the collection of points $q_v \in \mathscr{R}^n$ such that

$$M(q_v) = p$$

and $q_v \notin s$. Then at least one coordinate of q_v has a nonzero imaginary part. Suppose

$$q_v = (a_1 + ib_1, \cdots, a_n + ib_n).$$

Let

$$\bar{q}_v = (a_1 - ib_1, \cdots, a_n - ib_n).$$

If $q_v \in E$, then $\bar{q}_v \in E$ because the functions $f_i(x_1, \cdots, x_n)$ are real-valued $(i = 1, \cdots, n)$. Now E has no limit points in s because if there were such a limit point q_0, then $M(q_0) = p$ by the continuity of $\mathfrak{M}$. Thus there would be a p-point in s which would not be isolated. But $J(p) \neq 0$ by hypothesis. Hence there exists $\varepsilon > 0$ such that the set

$$F = \left[(a_1 + ib_1, \cdots, a_n + ib_n) \,/\, (a_1, \cdots, a_n) \in s, \sum_{i=1}^{n} |b_i| < \varepsilon\right]$$

contains the p-points on s but no other p-points and also such that the set $S^{(2)} - F$ contains no p-points on its boundary. Now from the definition of local degree, it follows that:

$$d[M^{(2)}, S^{(2)}, \bar{0}] = r + d[M^{(2)}, S^{(2)} - F, \bar{0}].$$

We complete the proof by showing that $d[M^{(2)}, S^{(2)} - F, \bar{0}]$ is an even number. Let p be a point in a neighborhood of $\bar{0}$ such that

$$d[M^{(2)}, S^{(2)} - F, \bar{0}] = d[M^{(2)}, S^{(2)} - F, p]$$

and such that p is not the image of a critical point of $M^{(2)}$. If

$$q_v \in [(M^{(2)})^{-1}(p)] \cap [S^{(2)} - F],$$

then

$$\bar{q}_v \in [(M^{(2)})^{-1}(p)] \cap [S^{(2)} - F].$$

Since each point has index $+1$, then $d[M^{(2)}, S^{(2)} - F, \bar{0}]$ is the number of points in $[(M^{(2)})^{-1}(p)] \cap [S^{(2)} - F]$ and hence is an even number.

Although we will not make use of it here, another possible method for computing the local degree is to use the Kronecker integral (Alexandroff and Hopf [**1**, pp. 465–567]). The order $v[f, \bar{K}, p]$ can be shown to be equal to an integral over $\bar{K}$, the Kronecker integral. (Indeed, historically the Kronecker integral precedes the definition of local degree we have developed here.) However, computing the integral is generally difficult.

10. **A reduction theorem and an in-the-large implicit function theorem.** We add two theorems which will be useful in applications. The first theorem is due to Leray and Schauder [**1**].

Let R^n be a Euclidean n-space regarded as a subset of R^{n+1}, an oriented Euclidean $(n + 1)$-space. Let y be a point in R^{n+1} such that $y \notin R^n$. The orientation of R^n is induced by the orientation of R^{n+1} in this way: the n-simplex $(x_0 \cdots x_n)$ is positively oriented if the simplex $(yx_0 \cdots x_n)$ in R^{n+1} (where $y \in R^{n+1}$ and $\notin R^n$) is positively oriented in R^{n+1}. Let $b \in R^n$ and ω_{n+1} be a bounded open set in R^{n+1} such that $\omega_{n+1} \cap R^n \neq \varnothing$. Then $\omega_n = \omega_{n+1} \cap R^n$ is a nonempty bounded open set in R^n and $\omega_n' = \omega_{n+1}' \cap R^n$ where ω_n' and ω_{n+1}' are the boundaries of ω_n and ω_{n+1}, respectively. Let F be a continuous mapping from $\bar{\omega}_{n+1} = \omega_{n+1} \cup \omega_{n+1}'$ into R^{n+1} such that for each $x \in \bar{\omega}_{n+1}$,

$$F(x) = x + g_n(x)$$

where $g_n(x) \in R^n$ and suppose $b \notin F(\omega_{n+1}')$. Then $f = F/R^n$ is such that $f(\bar{\omega}_n) \subset R^n$ and $b \notin f(\omega_n')$. Thus both $d[f, \bar{\omega}_n, b]$ and $d[F, \bar{\omega}_{n+1}, b]$ are defined.

(10.1) Reduction Theorem. $d[f, \bar{\omega}_n, b] = d[F, \bar{\omega}_{n+1}, b]$.

Proof. Approximate F by a continuous simplicial mapping S_{n+1} whose $(n + 1)$-simplexes each have one n-face which is parallel to R^n and such that none of the vertices of the $(n + 1)$-simplexes is in R^n. Let $(x_0x_1 \cdots x_{n+1})$ be such an $(n + 1)$-simplex and let $(x_1 \cdots x_{n+1})$ be the n-simplex which is parallel to R^n. Suppose $\overline{(x_0x_1 \cdots x_{n+1})} \cap R^n \neq \varnothing$. The line segment $\overline{x_0x_i}$ $(i = 1, \cdots, n + 1)$ intersects R^n in a single point y_i and $(y_1y_2 \cdots y_{n+1})$ is an n-simplex in R^n. Now $S_n = S_{n+1}/R^n$ is a continuous simplicial mapping which approximates f as closely as S_{n+1} approximates F.

The proof is completed by showing that

$$\text{(10.1a)} \qquad i[S_n(y_1y_2 \cdots y_{n+1}), b] = i[S_{n+1}(x_0x_1 \cdots x_{n+1}), b].$$

We need only consider the case: $S_{n+1}(x_0x_1 \cdots x_{n+1})$ is an $(n + 1)$-simplex and b is a point in the interior of $\overline{S_{n+1}(x_0x_1 \cdots x_{n+1})}$. Thus $S_n(y_1 \cdots y_{n+1})$ is an n-simplex and b is in the interior of $\overline{S_n(y_1 \cdots y_{n+1})}$. Since S_{n+1} moves all points parallel to R^n, the orientation of $S_{n+1}(x_0x_1 \cdots x_{n+1})$ in R^{n+1} is positive or negative according as the orientation of $S_n(y_1 \cdots y_{n+1})$ is positive or negative in R^n. Hence (10.1a) holds. ∎

(10.2) An In-The-Large Implicit Function Theorem. *Let $F_\mu(y)$ be a mapping from $R^n \times [-\varepsilon, \varepsilon]$ into R^n which is differentiable as a function of y and continuous in μ. Suppose $d[F_0, \bar{D}, \bar{0}]$ is defined where $\bar{0}$ is the origin of R^n, D is a bounded open set, and $|d[F_0, \bar{D}, \bar{0}]| = m > 0$. Then there is a $\delta > 0$ such that for all points x in $N_\delta(\bar{0})$ except for a set of n-measure zero, the equation*

$$F_\mu(y) - x = 0$$

has at least m distinct solutions $y_1(\mu), \cdots, y_q(\mu)$ where $q \geqq m$ and each of the solutions is a continuous function of μ for some neighborhood (on the μ-axis) of $\mu = 0$. The range of each of the functions $y_1(\mu), \cdots, y_q(\mu)$ is contained *in D. Also these solutions do not have any common values, i.e., if μ_i is a point in the domain of $y_i(\mu)$ and μ_j is a point in the domain of $y_j(\mu)$, then*

$$y_i(\mu_i) \neq y_j(\mu_j).$$

PROOF. Since $d[F_0, D, \bar{0}]$ is defined, there is a neighborhood

$$N_{\varepsilon_1}(\bar{0}) = [x \,/\, |x| < \varepsilon_1]$$

such that, except for a set of n-dimensional measure zero, if $x \in N_{\varepsilon_1}(\bar{0})$, then x is not the image of a critical point of $F_0(y)$ and $d[F_0, \bar{D}, x]$ is defined and is equal to $d[F_0, \bar{D}, \bar{0}]$. Let x_0 be such a point which is not the image of a critical point. Then $F_0^{-1}(x_0)$ is a finite set of points $y_1, \cdots, y_q$ where $q = m$. The Jacobian of $F_0(y)$ is nonzero at each of the points $y_1, \cdots, y_q$. Let $N(y_1), \cdots, N(y_r)$ be neighborhoods of $y_1, \cdots, y_r$, respectively, such that the Jacobian of $F_0(y)$ is nonzero in each $N(y_i)$, $i = 1, \cdots, r$, and such that the neighborhoods $N(y_1), \cdots, N(y_r)$ are pairwise disjoint. Use these neighborhoods and apply the usual implicit function theorem to each of the initial solutions $\mu = 0, y = y_1, \cdots, \mu = 0, y = y_r$ of the equation:

$$F_\mu(y) - x_0 = \bar{0}. \blacksquare$$

11. **A proof of the fixed point theorem.** It is easy to show that the fixed point theorem follows from the corresponding statement for a solid unit sphere S with center $\bar{0}$ in R^n. So we prove this latter statement.

(11.1) THEOREM. *If f is a continuous mapping of S into itself, then there is a point $x \in S$ such that $f(x) = x$.*

PROOF. First we may suppose that there is no point $y \in S'$ such that $y - f(y) = \bar{0}$. Otherwise the theorem would be true. Let $y \in S'$. Then $y - f(y) \neq \bar{0}$ by the supposition. If $0 \leqq t < 1$, then $y - tf(y) \neq \bar{0}$ because $y \in S'$ and $|tf(y)| < 1$ so that $tf(y) \notin S'$. Thus by the Invariance under Homotopy Theorem (6.4),

$$d[I, S, \bar{0}] = d[I - f, S, \bar{0}]$$

where I is the identity mapping. But $d[I, S, \bar{0}] = 1$. Hence by the Existence Theorem (6.6), the equation

$$x - f(x) = \bar{0}$$

has at least one solution x in the interior of S. $\blacksquare$

A special case of the fixed point theorem which follows at once from Theorem (11.1) is:

(11.2) COROLLARY. *If K is a convex set and f is a continuous mapping of K into itself, then there is a point $x \in K$ such that $f(x) = x$.*

12. **The index of a fixed point.** In the preceding section, we proved that the mapping f has a fixed point by showing that $d[I - f, S, \bar{0}] \neq 0$. Extend-

ing this viewpoint, we may define the fixed point index of an isolated fixed point p of a mapping f. (The fixed point p is isolated if there is a neighborhood of p such that the only fixed point of f in the neighborhood is p.) The index of a fixed point p is defined to be $d[I - f, S, \overline{0}]$ where S is a sphere with center $\overline{0}$ and radius r such that p is the only fixed point of f in S.

13. The index of a vector field. It is convenient in some applications to work with the concept of the index of a vector field rather than the local degree. The index is easily defined in terms of local degree in this way: let $B^n = g(\sigma^n)$ where σ^n is the solid n-sphere and g is a topological mapping.

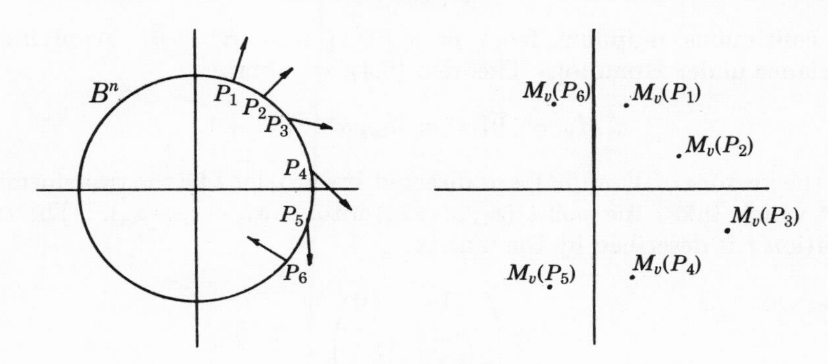

FIGURE 9

DEFINITION. A continuous mapping V from B^n into the space of n-vectors is a *vector field on* B^n.

DEFINITION. If $V(x_0)$ is the zero vector for some $x_0 \in B^n$, then x_0 is a *critical point* of the vector field V.

If for each $x \in B^n$, the vector $V(x)$ is a vector $\overline{\overline{0a}}$, i.e., with the head of the vector at point a, then the mapping

$$M_V : x \to a$$

is a continuous mapping from B^n into R^n. (See Figure 9.) If the vector field V has no critical points on $(B^n)'$, the boundary of B^n, then $d[M_V, B^n, \overline{0}]$ is defined.

DEFINITION. The *index of vector field* $V(x)$ relative to $(B^n)'$ is $d[M_V, B^n, 0]$.

If $n = 2$, the index is the algebraic number of rotations through angle 2π made by the vector as one proceeds around $(B^2)'$ in the counterclockwise direction. (A rotation of the vector in the counterclockwise [clockwise] direction is counted as a positive [negative] rotation.)

(13.1) THEOREM. *If B^n is the solid n-sphere σ^n and if V is a vector field on σ^n such that every vector on $(\sigma^n)'$ is directed outward [inward] the index of V relative to $(\sigma^n)'$ is* $+1[(-1)^n]$.

PROOF. If all the vectors of V are directed outward, define the vector field W on σ^n as follows: for each $x \in \sigma^n$ such that $x \neq \bar{0}$ let $W(x)$ be the vector $\overline{\bar{0}a}$. If $x = \bar{0}$, let $W(x)$ be the zero vector. Then M_W is the identity mapping and $d[M_W, \sigma^n, \bar{0}] = +1$. Now define the continuous family of vector fields for $t \in [0, 1]$:

$$(1 - t)W(x) + tV(x).$$

Then

$$M_{(1-t)W+tV}$$

is a continuous mapping from $\sigma^n \times [0, 1]$ into $R^n - \bar{0}$. Applying the Invariance under Homotopy Theorem (6.4), we obtain:

$$d[M_V, \sigma^n, \bar{0}] = d[M_W, \sigma^n, \bar{0}] = +1.$$

If all the vectors of V on $(\sigma^n)'$ are directed inward, let I be the transformation of R^n which takes the point $(x_1, \cdots, x_n)$ into $(-x_1, \cdots, -x_n)$. The transformation I is described by the matrix

$$\begin{pmatrix} -1 & & 0 \\ & \ddots & \\ 0 & & -1 \end{pmatrix}$$

and $d[I, \sigma^n, \bar{0}] = (-1)^n$. Now IM_V is the mapping corresponding to a vector field all of whose vectors on $(\sigma^n)'$ are directed outward. By the first part of the proof,

$$d[IM_v, \sigma^n, \bar{0}] = +1.$$

Hence by the Product Theorem (8.1),

$$(-1)^n d[M_v, \sigma^n, \bar{0}] = +1$$

or

$$d[M_v, \sigma^n, \bar{0}] = (-1)^n. \blacksquare$$

14. Generalizations. The theory of fixed points and local degree described in this chapter is only for mappings whose domains and ranges are subsets of Euclidean space. In Chapter III, the theory will be extended to certain mappings of which the domains and ranges are subsets of linear normed spaces. This theory will be sufficient for the applications to analysis, but there are far wider generalizations. The most important is the Lefschetz fixed point theory (see Alexandroff and Hopf [**1**], and Lefschetz [**2**; **3**]). Using combinatorial concepts (in particular, the Lefschetz Number) Lefschetz obtained a fixed point theorem for mappings in a wide class of abstract topological spaces. The fixed point theorems in Chapters I and III may be obtained as very special cases of this general theorem. A fixed point index for mappings in abstract spaces has been defined by Browder [**1**] and Bourgin [**1**]. A generalization in another direction has been carried out by Kakutani [**2**] who studied mappings from points to sets.

CHAPTER II

Applications to Ordinary Differential Equations

We will apply the techniques described in Chapter I to study the existence and stability of periodic and almost periodic solutions of nonlinear systems of ordinary differential equations. Since our object is to provide only an introduction to this kind of study, we give a fairly self-contained account of some of the simplest applications to quasilinear systems, but a much briefer description of the more difficult applications to systems with large nonlinearities. Also, the discussion is not complete. For example, we omit the work of Cesari [**3**] on quasilinear systems; in Chapter IV, however, we describe the extended form of Cesari's method to systems with large nonlinearities.

1. **Some existence theorems for differential equations.** Throughout this chapter, we will use vector notation and apply certain existence theorems, and we begin with a discussion of these. We will study systems of the form

$$\frac{dx_i}{dt} = f_i(x_1, \cdots, x_n; t) \qquad (i = 1, \cdots, n) \tag{1.1}$$

where the f_i are functions of $x_1, \cdots, x_n$ and t. Such a system is called an n-dimensional first-order system. An nth-order differential equation

$$\frac{d^n x}{dt^n} = f\left(x, \frac{dx}{dt}, \cdots, \frac{d^{n-1}x}{dt^{n-1}}; t\right) \tag{1.2}$$

can be regarded as an n-dimensional first-order system if we write $x = x_1$ and use the notation

$$x_{i+1} = \frac{dx_i}{dt} = \dot{x}_i \qquad (i = 1, \cdots, n - 1).$$

Then equation (1.2) becomes the n-dimensional first-order system:

$$\begin{aligned} \dot{x}_1 &= x_2 \\ \dot{x}_2 &= x_3 \\ &\cdots \\ \dot{x}_{n-1} &= x_n \\ \dot{x}_n &= f(x_1, x_2, \cdots, x_n; t). \end{aligned}$$

Frequently it will be convenient to write (1.1) in the vector form:

$$\dot{x} = f(x; t),$$

where x and f are vectors.

The subject of existence theorems for solutions of (1.1) has been treated at length by numerous writers. We will prove the simplest version of the basic existence theorem and a theorem on continuation of solutions. We will merely state and explain the significance of the other theorems we need.

DEFINITION. A *domain* is a connected open set in R^n.

DEFINITION. A *convex set* E in R^n is a set such that if p, q are points in E, then the line segment joining p and q is contained in E.

DEFINITION. Let $f(x; t)$ be a real n-vector function (i.e., a function with range in R^n) defined on a domain D in (x, t)-space where x is a real n-vector and t is real. Suppose there is a positive constant k such that if $(x^{(1)}, t)$ and $(x^{(2)}, t)$ are points in D, then

$$|f(x^{(1)}; t) - f(x^{(2)}; t)| \leqq k|x^{(1)} - x^{(2)}|.$$

Then *f satisfies a Lipschitz condition with respect to x in D* and k is a *Lipschitz constant of f*.

For example, if the components of f are denoted by $f_1, \cdots, f_n$ and if $\partial f_i / \partial x_j$ $(i, j = 1, \cdots, n)$ exists and is bounded in D, a convex domain, then by the mean value theorem, the function f satisfies a Lipschitz condition with respect to x in D. Note that if f satisfies a Lipschitz condition in x, then f is uniformly continuous in x but f may not even be continuous in t.

Now suppose $f(x; t)$ is an n-vector function, where x is an n-vector and t is a real variable, and f is defined on the set

$$R = \{(x, t) \; / \; |x - x^{(0)}| \leqq a, \;\; |t - t_0| \leqq b\}$$

where $x^{(0)}$ and t_0 are fixed and a and b are positive numbers. Suppose f is continuous on R and let $M = \max|f|$ on R. Let α be the minimum of a and b/M.

(1.3) BASIC EXISTENCE THEOREM. *Suppose $f(x; t)$ is continuous on a domain D such that $D \supset R$ and suppose f satisfies a Lipschitz condition with respect to x in D. Then the n-dimensional system*

$$\dot{x} = f(x; t) \tag{1.4}$$

has a unique solution $x(t, x^{(0)}, t_0)$ such that

$$x(t_0, x^{(0)}, t_0) = x^{(0)}.$$

Solution $x(t, x^{(0)}, t_0)$ is defined for all $t \in [t_0 - \alpha, t_0 + \alpha]$ and

$$|x(t, x^{(0)}, t_0) - x^{(0)}| \leqq M\alpha$$

for all $t \in [t_0 - \alpha, t_0 + \alpha]$.

Roughly speaking, the theorem says that for any pair $(x^{(0)}, t_0) \in D$, there is a unique solution of (1.4) which passes through $(x^{(0)}, t_0)$ or, more precisely, which satisfies the initial condition specified by $(x^{(0)}, t_0)$. The Lipschitz condition on f guarantees the uniqueness of the solution. Weaker hypotheses than the Lipschitz condition imply uniqueness of solution, but the Lipschitz condition, because of its simplicity, is very frequently used.

A striking feature of the basic existence theorem above which holds generally in existence theorems for ordinary differential equations is that the solution $x(t, x^{(0)}, t_0)$ is only defined *locally*. That is, the conclusion of the theorem says only that $x(t, x^{(0)}, t_0)$ is defined on $[t_0 - \alpha, t_0 + \alpha]$ and this may be a very small interval. That such a localness condition is inevitable is shown by the example:

$$\dot{x} = x^2$$

where x is a real variable. A solution of this equation which satisfies the initial condition $t_0 = 1$, $x^{(0)} = -1$ is $x(t) = -1/t$. This solution does not exist at $t = 0$ although since $f(x, t) = x^2$ is independent of t, function f can certainly be regarded as a well-behaved function at $t = 0$.

Proof of the basic existence theorem for the one-dimensional case: notice first that $x(t, x^{(0)}, t_0)$, continuous in t for $t \in [t_0 - \alpha, t_0 + \alpha]$, is a solution of (1.4) for $t \in [t_0 - \alpha, t_0 + \alpha]$, i.e.,

$$\dot{x}(t, x^{(0)}, t_0) = f[x(t, x^{(0)}, t_0); t]$$

if and only if

$$x(t, x^{(0)}, t_0) = x^{(0)} + \int_{t_0}^{t} f[x(s, x^{(0)}, t_0); s]\, ds. \tag{1.5}$$

We prove the theorem by showing that the sequence

$$x_0(t) \equiv x^{(0)},$$

$$x_{n+1}(t) = x^{(0)} + \int_{t_0}^{t} f[x_n(s); s]\, ds \qquad (n = 1, 2, \cdots)$$

converges to a solution of (1.5). The proof for the interval $[t_0, t_0 + \alpha]$ requires four steps. (Analogous arguments can be made for the interval $[t_0 - \alpha, t_0]$.)

Step 1. For each n, the function $x_n(t)$ is defined and has a continuous first derivative on $[t_0, t_0 + \alpha]$ and

$$|x_k(t) - x^{(0)}| \leqq M(t - t_0)$$

for $t \in [t_0, t_0 + \alpha]$.

The statement of this step is clearly true for $x_0(t)$ and we assume it is true for $x_n(t)$. Then the statement of the step for $x_{n+1}(t)$ follows from the standard theorem about differentiation of an integral and from the usual estimate for an upper bound of the absolute value of an integral.

Step 2. The sequence $\{x_n(t)\}$ converges uniformly on $[t_0, t_0 + \alpha]$ to a function $x(t)$.

Let k be a Lipschitz constant of f and let

$$d_n(t) = |x_{n+1}(t) - x_n(t)|$$

for $t \in [t_0, t_0 + \alpha]$. Then for all n,

$$\begin{aligned} d_{n+1}(t) &= \left| \int_{t_0}^{t} \{f[x_{n+1}(s); s] - f[x_n(s); s]\} \, ds \right| \\ &\leqq \int_{t_0}^{t} |f[x_{n+1}(s); s] - f[x_n(s); s]| \, ds \\ &\leqq k \int_{t_0}^{t} |x_{n+1}(s) - x_{n-1}(s)| \, ds = k \int_{t_0}^{t} d_n(s) \, ds. \end{aligned}$$

Now by induction, we obtain an estimate for $d_n(t)$. First:

$$d_0(t) = |x_1(t) - x_0(t)| \leqq M(t - t_0).$$

Suppose that for $t \in [t_0, t_0 + \alpha]$

$$d_n(t) \leqq \frac{M}{k} \frac{k^{n+1}(t - t_0)^{n+1}}{(n + 1)!}.$$

Then

$$\begin{aligned} d_{n+1}(t) &\leqq k \int_{t_0}^{t} d_n(s) \, ds \\ &\leqq k \frac{M}{k} \frac{k^{n+1}}{(n + 1)!} \int_{t_0}^{t} (s - t_0)^{n+1} \, ds \\ &= \frac{M}{k} \frac{k^{n+2}}{(n + 2)!} (t - t_0)^{n+2}. \end{aligned}$$

With this estimate, we have for $t \in [t_0, t_0 + \alpha]$

$$\begin{aligned} \sum_{n=0}^{\infty} d_n(t) &\leqq \frac{M}{k} \sum_{n=0}^{\infty} \frac{k^{n+1}(t - t_0)^{n+1}}{(n + 1)!} = \frac{M}{k} e^{k(t - t_0)} \\ &\leqq \frac{M}{k} e^{k\alpha}. \end{aligned}$$

Therefore by the Weierstrass M-test,

$$x_0(t) + \sum_{n=0}^{\infty} [x_{n+1}(t) - x_n(t)]$$

converges absolutely and uniformly on $[t_0, t_0 + \alpha]$ to a function $x(t)$. Since the sequence of partial sums, $\{x_n(t)\}$, converges uniformly to $x(t)$, then $x(t)$ is continuous on $[t_0, t_0 + \alpha]$.

Step 3. $x(t)$ is a solution of (1.5).

Since for $t \in [t_0, t_0 + \alpha]$ and for all n,

$$|x_n(t) - x^{(0)}| \leqq M(t - t_0),$$

then

$$|x(t) - x^{(0)}| \leqq M(t - t_0).$$

Hence for all $t \in [t_0, t_0 + \alpha]$, the function $f[x(t); t]$ is defined and

$$\left| \int_{t_0}^{t} \{f[x(s); s] - f[x_n(s); s]\} \, ds \right| \leqq k \int_{t_0}^{t} |x(s) - x_n(s)| \, ds. \tag{1.6}$$

Since $x_n(s)$ converges uniformly to $x(s)$ for $s \in [t_0, t_0 + \alpha]$, then

$$\lim_{n \to \infty} \int_{t_0}^{t} |x(s) - x_n(s)| \, ds = 0$$

uniformly in t. By (1.6),

$$\lim_{n \to \infty} \int_{t_0}^{t} f[x_n(s); s] \, ds = \int_{t_0}^{t} f[x(s); s] \, ds$$

uniformly in t for $t \in [t_0, t_0 + \alpha]$. This permits us to take the limit as $n \to \infty$ on both sides of the defining equation

$$x_{n+1}(t) = x^{(0)} + \int_{t_0}^{t} f[x_n(s); s] \, ds.$$

We obtain the desired conclusion.

Step 4. $x(t)$ is a unique solution of (1.5).

Suppose there exist continuous solutions $x(t)$ and $x^{(1)}(t)$ of (1.5).

Since $|x(t) - x^{(1)}(t)|$ is continuous on $[t_0, t_0 + \alpha]$, there is a positive constant A such that for $t \in [t_0, t_0 + \alpha]$,

$$|x(t) - x^{(1)}(t)| \leqq A.$$

But then

$$|x(t) - x^{(1)}(t)| \leqq k \int_{t_0}^{t} |x(s) - x^{(1)}(s)| \, ds \leqq kA(t - t_0).$$

Using this last estimate, we obtain by the same kind of procedure:

$$|x(t) - x^{(1)}(t)| \leqq \tfrac{1}{2} k^2 A (t - t_0)^2.$$

Continuing by induction, we obtain:

$$|x(t) - x^{(1)}(t)| \leqq \frac{A}{m!} [k(t - t_0)]^m,$$

the $(m + 1)$st term in the series representation for $Ae^{k(t - t_0)}$ for all m and all $t \in [t_0, t_0 + \alpha]$. Therefore,

$$x(t) \equiv x^{(1)}(t)$$

for $t \in [t_0, t_0 + \alpha]$. ∎

The localness condition in the conclusion of the Basic Existence Theorem (1.3) does not preclude the possibility that the solution can be "continued." That is, there may exist a solution $x^{(1)}(t)$ of (1.4) such that the domain of the function $x^{(1)}(t)$ contains properly the domain of solution $x(t)$ and such that $x^{(1)}(t) \equiv x(t)$ on the domain of $x(t)$. Solution $x^{(1)}(t)$ is called a *continuation* of solution $x(t)$.

(1.7) CONTINUATION THEOREM. *Let D be a bounded domain in (x, t)-space such that $f(x, t)$ is continuous in D and $f(x, t)$ satisfies a Lipschitz condition with respect to x in some neighborhood of each point in D. Suppose that the solution $x(t)$ of*

$$\dot{x} = f(x, t) \tag{1.4}$$

which passes through (x_0, t_0) can be defined for t greater than t_0 only in the interval $[t_0, t_0 + m)$. Let $\delta(t)$ be the distance from the point $(t, x(t))$ to the boundary of D, i.e.,

$$\delta(t) = \operatorname*{glb}_{p \in D'} |(t, x(t)) - p|,$$

where D' is the set of boundary points of D. (Since $x(t)$ is continuous, the glb *is a minimum.) Then*

$$\lim_{t \to (t_0 + m)} \delta(t) = 0.$$

If D is unbounded and the other hypotheses are the same,

(i) $\lim_{t \to (t_0 + m)} |x(t)| = \infty$ *or*
(ii) $\lim_{t \to (t_0 + m)} \delta(t) = 0.$

PROOF. Let S be the open set in D which consists of those points of D which have distance greater than ε_1 from D' where ε_1 is a fixed positive number. Suppose the curve in the (x, t)-plane described by $(t, x(t))$ has points in S as t approaches $t_0 + m$. More precisely, suppose: if $\varepsilon > 0$, then there is a point t_ε such that

$$t_0 + m - \varepsilon < t_\varepsilon < t_0 + m$$

and such that $(t_\varepsilon, x(t_\varepsilon)) \in S$. Then there is a sequence $\{t_j\}$ such that

$$t_j < t_{j+1} < t_0 + m$$

for all j and $\lim_{j \to \infty} t_j = t_0 + m$ and $(t_j, x(t_j)) \in S$ for all j. Since S is bounded, the set $(t_j, x(t_j))$ has a cluster point $\tilde{P} = (\tilde{x}, \tilde{t})$ in $S \cup S'$. Since $\tilde{P} \in D$, there is a rectangle $R \subset D$ with edges which are segments of the lines $t = \tilde{t} \pm c$, $x = \tilde{x} \pm d$ where c, d are positive constants. Suppose $|f(x, t)| \leqq M^{(1)}$ for all $(x, t) \in R$. We may assume $c < d/4M^{(1)}$. Let

$$R_1 = \{(x, t) \,/\, \tilde{t} - c \leqq t \leqq \tilde{t},\ |x - \tilde{x}| \leqq d/2\}.$$

Since $\lim_{j\to\infty} t_j = t_0 + m$, then $\tilde{t} = t_0 + m$. Therefore $t_j < \tilde{t}$ for all j; and given j_0, there is a $j > j_0$ such that $(t_j, x(t_j)) \in R_1$. Let $P_n = (t_n, x(t_n))$ be a fixed point in R_1 and apply the Basic Existence Theorem (1.3) with the initial values $(t_n, x(t_n))$. Then a solution of (1.7) is defined in the interval $[t_n, t_n + \alpha]$ where

$$\alpha = \text{minimum}\left[\tilde{t} + c - t_n, \frac{d}{2M^{(1)}}\right].$$

But $\tilde{t} + c - t_n \leqq 2c$ and $d/2M^{(1)} > 2c$ since $c < d/4M^{(1)}$. Hence $\alpha = \tilde{t} + c - t_n$. Thus the solution obtained is defined for t such that:

$$t_n \leqq t \leqq \tilde{t} + c.$$

Since $\tilde{t} = t_0 + m$, this contradicts the hypothesis that no continuation of $x(t)$ is defined at $t_0 + m$. As the ε_1 used at the beginning of the proof to define S is arbitrary, this completes the proof of the first statement in the theorem.

For the proof of the second statement, suppose that conclusion (i) does not hold. Then $|x(t)|$ remains bounded and there is a bounded domain D_1 such that $(t, x(t)) \in D_1$ for $t_0 < t \leqq t_0 + m$. Then we may apply the first statement of the theorem and obtain conclusion (ii). ∎

Now we state two extensions of the basic existence theorem: for proofs see Coddington and Levinson [**1**], Hurewicz [**1**] or Lefschetz [**4**].

(1.8) Existence Theorem for a System with a Parameter. *Let $f(x, t, \mu)$ be an n-vector function of n-vector x and real t and real μ. Suppose f is continuous on a domain D_μ in (x, t, μ)-space where D_μ is described by:*

(i) *$(x, t) \in D$, a domain in (x, t)-space;*
(ii) *$\mu \in [\mu \,/\, |\mu - \mu_0| < c]$ where μ_0 is constant and c is a positive constant.*

Suppose f satisfies a Lipschitz condition in x on D_μ and the Lipschitz constant is independent of μ. For $\mu = \mu_0$, suppose $x(t)$ is a solution of

$$\dot{x} = f(x, t, \mu) \tag{1.9}$$

for $t \in [a, b]$. Then there is a positive number δ such that if

$$\begin{gathered} t_0 \in \varepsilon(a, b), \\ |x_0 - x(t_0)| + |\mu - \mu_0| < \delta \end{gathered} \tag{1.10}$$

then there is a unique solution $X(t, \mu, t_0, x_0)$ of (1.9) defined for $t \in [a, b]$ and such that

$$X(t, \mu, t_0, x_0) = x_0$$

if $t = t_0$. Solution X is continuous in (t, μ, t_0, x_0) for $t \in (a, b)$ and for (μ, t_0, x_0) satisfying (1.10) *above.*

Notice that in this theorem we assume a solution given if parameter μ has a fixed value μ_0. The theorem says that for each μ sufficiently close to μ_0, equation (1.9) has a unique solution close to the given solution and with the same domain as the given solution. Thus in this theorem the localness condition is on the parameter μ.

A question about solutions which frequently occurs is: if $f(x, t)$ has certain differentiability properties, does the solution $x(t)$ have the same differentiability properties? The specific answers to this question which we need are given by the following theorem:

(1.11) DIFFERENTIABILITY THEOREM. *If $f(x, t, \mu)$ has continuous first [second] derivatives in x, t, and μ, then solution $X(t, \mu, t_0, x_0)$ of (1.9) has continuous first [second] derivatives in μ, t_0, x_0 and a continuous second [third] derivative in t.*

2. **Linear systems.** For our study of quasilinear systems, we will need some facts concerning linear systems. For proofs, see Coddington and Levinson [**1**]. Let

$$\dot{x} = A(t)x \tag{2.1}$$

be an n-dimensional linear homogeneous system in which the elements of the matrix $A(t)$ have continuous first derivatives in some interval (a, b), finite or infinite. (The results we state in this section hold with weaker conditions on $A(t)$ but we will need the above hypothesis on $A(t)$ later in our work.) System (2.1) has exactly n linearly independent solutions which are defined in the same interval (a, b). An $n \times n$ matrix whose columns are n linearly independent solutions of (2.1) is called a *fundamental matrix* of (2.1). If $M(t)$ is a fundamental matrix, then each solution of (2.1) can be written in the form $M(t)c$ where c is a constant n-vector. A fundamental matrix (its elements are functions of t) is nonsingular for all values of t. If A is a constant matrix, then

$$e^{tA} = E + tA + \frac{t^2A^2}{2!} + \cdots + \frac{t^nA^n}{n!} + \cdots$$

is a fundamental matrix of $\dot{x} = Ax$. The solution $x(t)$ of (2.1) which satisfies the initial condition

$$x(t_0) = x^{(0)}$$

is

$$x(t) = e^{(t-t_0)A}x^{(0)}.$$

For linear inhomogeneous systems, we have the useful *variation of constants formula*. Suppose $M(t)$ is a fundamental matrix of (2.1) and $x_h(t)$ is the solution of (2.1) which satisfies the initial condition:

$$x_h(t_0) = x_0.$$

Let $b(t)$ be an n-vector function continuous in t. Then the solution $x(t)$ of

$$\dot{x} = A(t)x + b(t)$$

which satisfies the initial condition $x(t_0) = x_0$ is

$$x(t) = x_h(t) + M(t) \int_{t_0}^{t} [M(s)]^{-1} b(s)\, ds. \tag{2.2}$$

3. Existence of periodic solutions of nonautonomous quasilinear systems. We study the existence and stability of periodic and almost periodic solutions of the n-dimensional system

$$\dot{x} = A(t)x + \mu f(x, t, \mu) + g(t) \tag{3.1}$$

where μ is a "small" real parameter. Because μ is small, equation (3.1) is often called a *quasilinear* system. We consider first the problem of periodic solutions of (3.1).

The hypotheses to be imposed on (3.1) are the following:

(H-1) The elements of the matrix $A(t)$ and the components of f and g are defined for all t and these functions have continuous second derivatives in t for all t.

(H-2) For all t, the components of f are defined for μ in an interval containing $\mu = 0$ and for all x in some domain D. The components of f have continuous second derivatives in μ and in the components of x.

(H-3) Matrix $A(t)$ and functions $f(x, t, \mu)$ and $g(t)$ have period T in variable t, i.e., for all t,

$$A(t + T) = A(t);$$
$$f(x, t + T, \mu) = f(x, t, \mu) \text{ for all } x, \mu;$$
$$g(t + T) = g(t).$$

The problem we study is: does equation (3.1) have solutions of period T if μ is sufficiently small?

First a word about the significance of this problem. The fact that μ is kept small might seem to imply that this is not a difficult problem. That is, one might argue that the linear equation obtained from (3.1) by setting μ equal to zero is a good approximation to (3.1) so that the periodic solutions of the linear equation would approximate the periodic solutions of (3.1). For certain circumstances (called the nonresonance case) this is a rough description of what actually occurs. However, the more interesting case, both from the physical and the mathematical points of view, is that in which the addition of the nonlinear term $\mu f(x, t, \mu)$ alters in a basic way the nature of the periodic solutions of (3.1). Starting with Poincaré [**1**] a great deal of work has been done on various aspects of this problem. A detailed discussion of much of this work, including applications, is given by Malkin [**1**].

We impose one further condition on (3.1). We require that $g(t) \not\equiv 0$ or that $f(x, t, \mu)$ or $A(t)$ be explicitly a function of t, i.e., we require that the right-hand side of (3.1) depend on t. Equation (3.1) is then said to be nonautonomous. If the right-hand side of (3.1) does not depend on t ((3.1) is then said to be autonomous) the condition that A, f, g have period T in t is trivially fulfilled. Indeed we may say that A, f, g have period T_0 in t where T_0 is an arbitrary positive number. In this case periodic solutions may be studied but the problem considered is quite different from the problem stated above. The crucial difference is that we can no longer require that the period of the solution be fixed as parameter μ is varied. The autonomous case will be studied in detail in § 8 of this chapter.

Now we proceed to a detailed study of (3.1). First setting $\mu = 0$, we obtain a linear system which can be solved (for example, by using the variation of constants formula) for all t. Applying the Existence Theorem for a System with a Parameter (1.8) and the Differentiability Theorem (1.11) we may conclude that if a positive ε is given, then there is a positive δ such that if $|\mu| < \delta$, then there is a solution $x(t, \mu, c)$ of (3.1) defined for $t \in [-\varepsilon, T + \varepsilon]$ such that $x(t, \mu, c)$ satisfies the initial condition $x(0, \mu, c) = c$ where c is a fixed n-vector and $x(t, \mu, c)$ has a continuous derivative in each component of c. We want to determine conditions under which this solution $x(t, \mu, c)$ has period T in t.

(3.2) THEOREM. *Solution $x(t, \mu, c)$ has period T in t if and only if*

$$x(T, \mu, c) - x(0, \mu, c) = 0. \tag{3.3}$$

PROOF. If $x(t, \mu, c)$ has period T in t, condition (3.3) is obviously satisfied. Suppose (3.3) holds. Let $y(t, \mu, c) = x(T + t, \mu, c)$. Then (3.3) becomes:

$$y(0, \mu, c) = x(0, \mu, c).$$

But $y(t, \mu, c)$ is a solution of (3.1) because

$$\begin{aligned}\dot{y}(t, \mu, c) = \dot{x}(T + t, \mu, c) &= A(T + t)x(T + t, \mu, c) \\ &\quad + \mu f[x(T + t, \mu, c), T + t, \mu] + g(T + t) \\ &= A(t)y(t, \mu, c) + \mu f[y(t, \mu, c), t, \mu] + g(t)\end{aligned}$$

by the definition of $y(t, \mu, c)$ and the periodicity in t of f, g, and A. Thus $y(t, \mu, c)$ and $x(t, \mu, c)$ are solutions of (3.1) which, because of the hypothesis, satisfy the same initial condition at $t = 0$. Hence by the uniqueness condition of the Basic Existence Theorem (1.3),

$$x(T + t, \mu, c) = y(t, \mu, c) = x(t, \mu, c)$$

for all t. ▌

Since $x(t, \mu, c)$ has period T in variable t then $x(t, \mu, c)$ can be extended periodically so that it is defined for all t.

Now we use the variation of constants formula (2.2) to rewrite the condition of periodicity (3.3). First let $M(t)$ be a fundamental matrix of

$$\dot{x} = A(t)x$$

such that $M(0) = E$, the identity matrix. Then

$$x(t, \mu, c) = M(t)c + M(t) \int_0^t [M(s)]^{-1}\{\mu f[x(s, \mu, c), s, \mu] + g(s)\}\, ds. \tag{3.3a}$$

Substituting (3.3a) in (3.3), we obtain:

$$(3.4)\quad [M(T) - M(0)]c + M(T) \int_0^T [M(s)]^{-1}\{\mu f[x(s, \mu, c), s, \mu] + g(s)\}\, ds = 0.$$

Since the general solution $x(t, \mu, c)$ is given, at least in a theoretical sense, by the Existence Theorem for a System with a Parameter (1.8), vector equation (3.4) is an equation in c and μ. If we can solve (3.4) for c as a function $c(\mu)$ of μ, then from the derivation of (3.4) it follows that the solution $x(t, \mu, c(\mu))$ given by Theorem (1.8) is a periodic solution of (3.1). Equation (3.4) can be regarded as a system of n equations in the n components of vector c in which case (3.4) is sometimes called the system of branching equations or bifurcation equations (or the Verzweigungsgleichungen). Thus the problem of finding solutions of (3.1) which have period T has become the problem of solving the branching equations (3.4) for c as a function of μ.

The nature of the solutions of the branching equations depends heavily on the rank of the matrix $[M(T) - M(0)]$. We first point out the relation of this rank to the properties of the linear part

$$\dot{x} = A(t)x$$

of the differential equation (3.1).

(3.5) THEOREM. *Matrix $[M(T) - M(0)]$ has rank $n - r$ if and only if*

$$\dot{x} = A(t)x$$

has exactly r linearly independent solutions of period T.

PROOF. Suppose $[M(T) - M(0)]$ has rank $n - r$. Then there exist exactly r constant linearly independent n-vectors $h_1, \cdots, h_r$ such that

$$[M(T) - M(0)]h_j = 0 \qquad j = 1, \cdots, r.$$

By the uniqueness of the solutions of $\dot{x} = A(t)x$,

$$[M(T + t) - M(t)]h_j = 0$$

for all t. Thus $M(t)h_j$ has period T $(j = 1, \cdots, r)$ and $M(t)h_1, \cdots, M(t)h_r$ are r linearly independent solutions of period T. So if R is the number of linearly independent solutions of period T, $R \geqq r$ or $n - R \leqq n - r$, the rank.

Now suppose

$$\dot{x} = A(t)x$$

has exactly R linearly independent solutions of period T. As $M(t)$ is a fundamental matrix, these R solutions can be written as $M(t)h_1, \cdots, M(t)h_R$ where $h_1, \cdots, h_R$ are linearly independent constant n-vectors. The periodicity implies:

$$[M(T) - M(0)]h_j = 0 \qquad (j = 1, \cdots, R).$$

So $M(T) - M(0)$ has rank less than or equal to $n - R$. ▮

We can divide the study of solving (3.3) into two cases.

Case I. Matrix $[M(T) - M(0)]$ is nonsingular. This is called the nonresonance case or the nondegenerate case. That $[M(T) - M(0)]$ is nonsingular implies that the implicit function theorem can be applied to solve (3.3) for c as a function of μ in a neighborhood of the initial solution:

$$\mu = 0,$$

$$c = -[M(T) - M(0)]^{-1}M(T) \int_0^T [M(s)]^{-1}g(s)\, ds.$$

This is because the determinant of $[M(T) - M(0)]$ is exactly the Jacobian which appears in the hypothesis of the implicit function theorem. Thus the existence of a unique periodic solution of (3.1) is established for sufficiently small μ.

This solution of the nonresonance case is due originally to Poincaré. The drawback to this result is that many of the most important physical problems described by systems of the form (3.1) are resonance problems. (See Malkin [**1**] and Minorsky [**2**].)

Case II. The resonance or degenerate case: matrix $[M(T) - M(0)]$ has rank $n - r$ and $n - r < n$. The number r is called the degree of degeneracy of the system. Theorem (3.5) shows why this is termed the resonance case: since

$$\dot{x} = A(t)x$$

has a solution of period T, then the frequency of the "forcing terms" f and g is the same as a "natural frequency" of the system. To study the solutions $c(\mu)$ of (3.4) in the resonance case, we will compute the local degree of the associated mapping and apply the Existence Theorem (6.6) and the In-the-large Implicit Function Theorem (10.2) of Chapter I.

Assume that $n - r$, the rank of $[M(T) - M(0)]$, is such that $0 < n - r < n$. This is the more complicated case and its treatment requires the use of some linear algebra. The reader who is interested only in applications can omit the next few paragraphs and go on at once to the degenerate case in which $n - r = 0$. This latter is the case which most frequently arises in applications and its study does not require any linear algebra.

Let E_r be the null space of $[M(T) - M(0)]$ and let E_{n-r} be the complementary space of E_r. Let $x_1, \cdots, x_n$ be a basis for R^n such that $x_1, \cdots, x_r$ is a basis for E_r and $x_{r+1}, \cdots, x_n$ a basis for E_{n-r}. Let P_r and P_{n-r} be the projections of R^n into E_r and E_{n-r}. Now there is a nonsingular matrix H such that

$$H[M(T) - M(0)] = P_{n-r}.$$

(Matrix P_{n-r} is called the Hermite canonical form of $[M(T) - M(0)]$. See MacDuffee [**1**, pp. 35–37].) We rewrite the branching equations (3.4) in terms of the basis $x_1, \cdots, x_n$ described above. Then multiplying (3.4) by matrix H, we obtain:

$$P_{n-r}c + \mu HF(c, \mu) + d = 0, \tag{3.6}$$

where

$$F(c, \mu) = M(T) \int_0^T [M(s)]^{-1}\{\mu f[x(s, \mu, c), s, \mu]\} \, ds$$

and

$$d = HM(T) \int_0^T [M(s)]^{-1} g(s) \, ds.$$

(3.7) Theorem. *A necessary condition that* (3.6) *can be solved for c as a continuous function of* μ *is that* $P_r d = 0$.

Proof. Multiply (3.6) by P_{n-r} and P_r. Noting that $P_{n-r}P_r = P_r P_{n-r} = 0$ and $P_{n-r}P_{n-r} = P_{n-r}$ we obtain:

$$P_{n-r}c + \mu P_{n-r}HF(c, \mu) = -P_{n-r}d, \tag{3.7a}$$

$$\mu P_r HF(c, \mu) = -P_r d. \tag{3.7b}$$

Solving (3.6) is equivalent to solving (3.7a) and (3.7b) simultaneously for $P_{n-r}c$ and $P_r c$. Suppose we can solve for c as a continuous function $c(\mu)$ for μ such that $|\mu| \leqq \delta$. Then the function $c(\mu)$ and hence also the function $P_r HF(c(\mu), \mu)$ are bounded for $|\mu| \leqq \delta$. If $|P_r d| > 0$, then for μ sufficiently small,

$$|\mu P_r HF(c(\mu), \mu)| < \tfrac{1}{2}|P_r d|.$$

So (3.7b) cannot hold. ▌

Throughout the remainder of our discussion, we assume that $P_r d = 0$.

The expressions on the left in (3.7a) and (3.7b) with $\mu = 0$ in (3.7a) describe a mapping of the form considered in the Reduction Theorem (10.1) of Chapter I. If we compute the local degree at the origin of the mapping of R^r into R^r given by

$$\mu P_r HF[-P_{n-r}d + P_r c, \mu],$$

then by the Reduction Theorem (10.1) of Chapter I, the local degree of the mapping of R^n into R^n described by the left sides of (3.7a) and (3.7b) has the same value if $\mu = 0$ in (3.7a) and hence also if μ is sufficiently small in (3.7a). We can then apply the Existence Theorem (6.6) and the In-the-large Implicit Function Theorem (10.2) of Chapter I to establish the existence of solutions of (3.7a) and (3.7b) or, equivalently, of (3.4). This in turn yields periodic solutions of (3.1).

Now assume that $n - r = 0$, i.e., that $M(T) - M(0)$ is the zero matrix. This is sometimes called the totally degenerate case. Since $M(0) = E$, the identity matrix, then (3.4) becomes:

$$\mu \int_0^T [M(s)]^{-1}\{f[x(s, \mu, c), s, \mu]\}\, ds + \int_0^T [M(s)]^{-1}\, g(s) ds = 0. \tag{3.4a}$$

If

$$\int_0^T [M(s)]^{-1} g(s) ds \neq 0,$$

then since

$$\int_0^T [M(s)]^{-1}\{f[x(s, \mu, c), s, \mu]\} ds$$

is continuous in μ and c and hence bounded if $|\mu| \leqq \delta$ and $|c| \leqq A$ where δ and A are positive constants, it follows that if μ is sufficiently small, then

$$\left| \mu \int_0^T [M(s)]^{-1}\{f[x(s, \mu, c), s, \mu]\}\, ds \right| < \frac{1}{2} \left| \int_0^T [M(s)]^{-1} g(s)\, ds \right| .$$

Thus if μ were sufficiently small, equation (3.4a) would not be solvable. We have proved:

(3.8) THEOREM. *If $n - r = 0$, a necessary condition that (3.4) can be solved for c as a continuous function of μ is that*

$$\int_0^T [M(s)]^{-1} g(s)\, ds = 0.$$

Equation (3.4) becomes in this case:

$$\mu \int_0^T [M(s)]^{-1}\{f[x(s, \mu, c), s, \mu]\}\, ds = 0. \tag{3.9}$$

Again we apply the Existence Theorem (6.6) and the In-the-large Implicit Function Theorem (10.2) of Chapter I to study the solutions of (3.8). For example, from Theorem (6.6), we obtain:

(3.10) THEOREM. *Suppose $M : c \to c'$ is the mapping defined by*

$$c' = \mu \int_0^T [M(s)]^{-1}\{f[x(s, \mu, c), s, \mu]\}\, ds$$

where μ has a fixed value. If $\bar{D}$ is the closure of a bounded open set such that $d[M, \bar{D}, \bar{0}] \neq 0$, then equation (3.9) has a solution c which is a point in $\bar{D}$ and hence (3.1) has at least one periodic solution.

At this stage, Theorem (3.10) hardly seems useful since we have not given any method for determining $d[M, \bar{D}, \bar{0}]$. As we shall see in § 6, this local degree can easily be computed in many cases by using the techniques of Chapter I. Of course, Theorems (6.6) and (10.2) of Chapter I will not give a complete solution of the problem of periodic solutions. If the local degree is zero or if it is not defined, we cannot obtain much information as we will show with examples in § 6 of this chapter. For such cases, the local degree theory is too coarse.

4. **Some stability theory.** Before applying the technique described above to some concrete problems in nonlinear oscillations, we show that if (3.1) is a 2-dimensional system, then use of the local degree yields not only an estimate of the number of periodic solutions but also information about their stability properties.

In applications, it is essential to know if the solution (equilibrium point, periodic solution or whatever) has some kind of stability. If the solution is unstable, it will not correspond to an observed phenomenon in the physical system described by the equation. So from the point of view of applications, stability is just as important as existence. Indeed in some applications such as certain problems in the differential equations of control systems, interest is centered mainly on the stability question. The stability problem is hard to solve because of the difficulty of obtaining a satisfactory criterion for stability which is practical enough to be applied in specific problems. The treatment here will be based upon the version of the Lyapunov theory in Lefschetz [**1**] and in Malkin [**2**]. The Lyapunov theory is widely used and has several advantages. However, it is not a final solution to the stability problem. For example, systems can be exhibited which are stable in the sense of Lyapunov and which are unstable from the practical point of view. See Malkin [**2**, pp. 176–178] and La Salle and Lefschetz [**1**, pp. 121–126].

It is convenient to start by studying the stability of critical points. Let the point $x^{(0)}$ be a critical point of the system,

$$\dot{x} = f(x, t), \tag{4.1}$$

i.e., suppose there is a real number t_0 such that for $t \geqq t_0$,

$$f(x^{(0)}, t) = 0.$$

Assume that the hypotheses of the Basic Existence Theorem (1.3) are satisfied except that the domain D contains the set

$$R = \{(x, t) \mid |x - x^{(0)}| \leqq a, t \geqq t_0\}$$

where a is a fixed positive number and t_0 is fixed.

Without limiting the generality, we may assume that $x^{(0)}$ is the origin which we denote by $\bar{0}$.

(4.2) Definition. The critical point $\bar{0}$ is *stable with respect to* (4.1) if, given $\varepsilon > 0$ and $t_1 \geqq t_0$, there exists $\eta > 0$ such that every solution $x(t)$ of (4.1) with $|x(t_1)| < \eta$ is defined for all $t \geqq t_1$ and $|x(t)| < \varepsilon$ for all $t \geqq t_1$.

Critical point $\bar{0}$ is *asymptotically stable* if $|x(t_1)| < \eta$ implies that $|x(t)| < \varepsilon$ for all $t \geqq t_1$ and $\lim_{t\to\infty} x(t) = 0$. If critical point $\bar{0}$ is not stable, then it is said to be *unstable.*

(Lefschetz [**4**] makes a finer analysis of stability by introducing the notion of conditional stability. Some of the results obtained here can be extended by using this notion.)

(4.3) Definition. The solution $x^{(0)}(t)$ of (4.1), defined for all $t \geqq t_0$, is *stable* if, given $\varepsilon > 0$ and $t_1 \geqq t_0$, there exists $\eta > 0$ such that every solution $x^{(1)}(t)$ of (4.1) with

$$|x^{(1)}(t_1) - x^{(0)}(t_1)| < \eta$$

is defined for all $t \geqq t_1$ and

$$|x^{(1)}(t) - x^{(0)}(t)| < \varepsilon$$

for all $t \geqq t_1$. If, in addition,

$$\lim_{t\to\infty} [x^{(1)}(t) - x^{(0)}(t)] = 0$$

then $x^{(0)}(t)$ is *asymptotically stable.* If solution $x^{(0)}(t)$ is not stable, then it is said to be *unstable.*

A weaker but often useful kind of stability is orbital stability. Let

$$N\{x^{(0)}(t), \eta\} = [x \,/\, \text{there is a } t_0 \text{ such that } |x - x^{(0)}(t_0)| < \eta].$$

(4.4) Definition. Solution $x^{(0)}(t)$ is *orbitally stable relative to a set* S of solutions if $\varepsilon > 0$ implies there exists $\eta > 0$ and $t_0 > 0$ such that if $x^{(1)}(t) \in S$ and $x^{(1)}(t_0) \in N\{x^{(0)}(t), \eta\}$, then $x^{(1)}(t) \in N\{x^{(0)}(t), \eta\}$ for all $t \geqq t_0$.

The other types of orbital stability, i.e., orbitally asymptotically stable and orbitally unstable, may be defined in an analogous way. Stability implies orbital stability, but there are examples of solutions which are orbitally stable but not stable. See Lefschetz [**4**, pp. 83–84].

Now we state several versions of a classical stability criterion due to Lyapunov.

Definition. Two $n \times n$ matrices C and D the elements of which are complex numbers are *similar* if there is a nonsingular $n \times n$ matrix P such that

$$D = PCP^{-1}.$$

(4.5) Theorem (Jordan Canonical Form). *Every $n \times n$ matrix A is similar to a matrix*

$$J = \begin{pmatrix} J_0 & & & \\ & J_1 & & \\ & & \ddots & \\ & & & J_s \end{pmatrix}$$

where the unwritten terms are zero and J_0 *is a diagonal matrix,*

$$J_0 = \begin{pmatrix} \lambda_1 & & & & \\ & \lambda_2 & & & \\ & & \cdot & & \\ & & & \cdot & \\ & & & & \lambda_q \end{pmatrix},$$

in which $\lambda_1, \cdots, \lambda_q$ *may or may not be distinct; and if* $i = 1, \cdots, S$,

$$J_i = \begin{pmatrix} \lambda_{q+i} & 1 & & & \\ & \lambda_{q+i} & \cdot & & \\ & & \cdot & \cdot & \\ & & & \cdot & 1 \\ & & & & \lambda_{q+i} \end{pmatrix}.$$

The numbers $\lambda_1, \lambda_2, \cdots, \lambda_{q+S}$ *are the characteristic roots of* A, *i.e., the roots of the characteristic polynomial*

$$\det (\lambda E - A),$$

where E *is the identity matrix, and the dimension of the square matrix* J_i *is the multiplicity of the characteristic root* λ_{q+i}.

(4.6) First Stability Theorem for a Critical Point. *In the n-dimensional system*

$$\dot{x} = Ax + q(x, t), \tag{4.7}$$

suppose A *is a constant matrix and suppose that in the set,*

$$E = [(x, t) \,/\, |x| \leqq R, t \geqq \tau],$$

where R *is a positive constant and* τ *is a constant, the function* $q(x, t)$ *is continuous and satisfies a Lipschitz condition in* x *and that*

$$q(x, t) = o(|x|),$$

i.e.,

$$\lim_{|x| \to 0} \frac{q(x, t)}{|x|} = 0$$

uniformly in t. *If the characteristic roots of* A *all have negative real parts, the origin* $\bar{0}$ *is asymptotically stable with respect to* (4.7). *If there is at least one characteristic root with a positive real part, the origin is unstable.*

A detailed discussion of this and other stability theorems can be found in Malkin [**2**] or Cesari [**2**].

Now suppose that $x(t)$ is a periodic solution of period T of the n-dimensional system

$$\dot{x} = f(x, t) \tag{4.8}$$

where f has period T in t and has continuous first derivatives in t for all $t \geqq t_0$ and in the components of x for all $|x| \leqq R$, a positive constant. Substitute the function

$$y(t) = x(t) + u(t)$$

for x in (4.8). Since $x(t)$ is a solution of (4.8), we obtain

$$\dot{u} = B(t)u + g(u, t) \tag{4.9}$$

where B and g have period T in t and for all $t \geqq t_0$

$$\lim_{u \to 0} \frac{g(u, t)}{|u|} = 0.$$

If B is a constant matrix, then we can apply the First Stability Theorem for a Critical Point (4.6) and obtain:

(4.10) First Stability Theorem for a Periodic Solution. *If B is a constant matrix and if the characteristic roots of B all have negative real parts, then solution $x(t)$ of* (4.8) *is asymptotically stable* (*and, therefore, also orbitally asymptotically stable*). *If there is a characteristic root of B with a positive real part, solution $x(t)$ is unstable.*

We also need a stability theorem for periodic solutions which is applicable if matrix B is not constant. For this purpose, we introduce the characteristic exponents of a linear system with periodic coefficients.

Let

$$\dot{x} = B(t)x \tag{4.11}$$

be a real linear n-dimensional system in which the matrix $B(t)$ has period T in t. Let $M(t)$ be a fundamental matrix of (4.11). Since $B(t)$ has period T, then $M(T + t)$ is also a fundamental matrix of (4.11). Hence there is a constant nonsingular matrix C such that

$$M(T + t) = M(t)C. \tag{4.12}$$

If $N(t)$ is another fundamental matrix of (4.11), there is a constant nonsingular matrix D such that

$$M(t) = N(t)D. \tag{4.13}$$

Substituting from (4.13) into (4.12), we have:

$$N(T + t)D = N(t)DC$$

or

$$N(T + t) = N(t)DCD^{-1}.$$

It is easy to prove that similar matrices have the same characteristic polynomial and hence the same characteristic roots. Thus the matrices C and DCD^{-1} have the same characteristic roots. Hence although matrix

C depends upon which fundamental matrix $M(t)$ we consider, the characteristic roots of matrix C are independent of $M(t)$ and depend only on system (4.11) itself. This justifies the following definition.

(4.14) DEFINITION. The characteristic roots $\mu_1, \cdots, \mu_n$ of matrix C are said to be the *characteristic exponents* of system (4.11).

Now the following results can be obtained. (See Lefschetz [**4**, pp. 144–147].) There is a matrix K (not unique) such that

$$C = e^{TK}.$$

The matrix

$$Z(t) = e^{Kt}[M(t)]^{-1}$$

is nonsingular and has period T. The transformation

$$y = Z(t)x$$

reduces (4.11) to the linear system with constant coefficients

$$\dot{y} = Ky. \tag{4.15}$$

The origin $\bar{0}$ is a critical point of systems (4.11) and (4.15) and has the same stability properties with respect to both systems. If the characteristic roots of K are $\lambda_1, \cdots, \lambda_n$ and if we denote the real part of λ_i by Re λ_i, then Re $\lambda_i < 0[>0]$ if and only if $|\mu_i| < 1[>1]$. Using these facts and the First Stability Theorem for a Critical Point (4.6), we obtain:

(4.16) SECOND STABILITY THEOREM FOR A CRITICAL POINT. *In the n-dimensional system*

$$\dot{x} = A(t)x + q(x, t), \tag{4.17}$$

if $A(t)$ has period T and if $q(x, t)$ satisfies the same hypotheses as in the First Stability Theorem (4.6), *then if the characteristic exponents μ_n of the system $x = A(t)x$ all have absolute value less than one, the origin $\bar{0}$ is asymptotically stable with respect to* (4.17). *If there is at least one μ_n with absolute value greater than one, then the origin is unstable.*

From the First Stability Theorem for a Periodic Solution (4.10), we obtain:

(4.18) SECOND STABILITY THEOREM FOR A PERIODIC SOLUTION. *If the characteristic exponents of the system*

$$\dot{u} = B(t)u$$

where $B(t)$ is the matrix in equation (4.9) *all have absolute value less than* 1, *then periodic solution $x(t)$ of* (4.8) *is asymptotically stable. If there is a characteristic exponent of $B(t)$ with absolute value greater than* 1, *then $x(t)$ is unstable.*

5. Stability of periodic solutions of nonautonomous quasilinear systems. Now we are ready to study the stability of the periodic solutions of (3.1). The existence of the periodic solutions is established by applying to the branching equations (3.4) either the implicit function theorem (in the nonresonance case) or the Existence Theorem (6.6) and the In-the-large Implicit Function Theorem (10.2) of Chapter I (in the resonance case).

The result of this section is, essentially, to show that the sign of the local degree yields information about the stability of the periodic solutions. This result is a rigorization and extension of the Andronov-Witt stability method in the nonautonomous case. (See Andronov and Witt [**1**], Andronov and Chaikin [**1**] and, especially, Stoker [**1**].)

Assume throughout our discussion of stability that (3.1) is 2-dimensional. Now suppose we have obtained a solution $c(\mu)$ of (3.4) for all sufficiently small μ. By Theorem (3.2) and the derivation of (3.4), solution $x(t, \mu, c(\mu))$ of (3.1) has period T in t. We study the question: is $x(t, \mu, c(\mu))$ stable?

First consider the nonresonance case. Let $c^{(0)}$ be a solution of (3.4) and consequently of

$$x(T, \mu, c) - x(0, \mu, c) = 0 \tag{3.3}$$

and let c be a point near $c^{(0)}$. We investigate the difference

$$x(T, \mu, c) - x(T, \mu, c^{(0)}). \tag{5.1}$$

Substituting from the variation of constants formula (3.3a) into (5.1) we obtain:

$$\begin{aligned} x(T, \mu, c) &- x(T, \mu, c^{(0)}) \\ &= M(T)c + M(T) \int_0^T [M(s)]^{-1}\{\mu f[x(s, \mu, c), s, \mu] + g(s)\}\, ds \\ &\quad - M(T)c^{(0)} - M(T) \int_0^T [M(s)]^{-1}\{\mu f[x(s, \mu, c^{(0)}), s, \mu] + g(s)\}\, ds \\ &= [M(T)](c - c^{(0)}) + \mu M(T) \int_0^T [M(s)]^{-1}\{f[x(s, \mu, c), s, \mu] \\ &\qquad - f[x(s, \mu, c^{(0)}), s, \mu]\}\, ds. \end{aligned}$$

Since $M(t)$ is a fundamental solution, $M(T)$ is nonsingular.

Because

$$x(0, \mu, c) = c$$

and

$$x(0, \mu, c^{(0)}) = c^{(0)},$$

if μ is sufficiently small and if the characteristic roots of $M(T)$ have absolute value less than 1, then $x(t, \mu, c^{(0)})$ is asymptotically stable. This follows from the definition (4.14) of characteristic exponents and the Second Stability Theorem for a Periodic Solution (4.18). Since $M(T)$ can be computed, we can therefore determine λ_1, λ_2 and thus determine the stability properties of $x(t, \mu, c^{(0)})$.

For the totally degenerate case, i.e., the case in which $M(T) - M(0)$ is the zero matrix, a different approach is used. The treatment is based on work of Lefschetz [**4**, pp. 301–303] and Cronin [**6**]. Suppose that in equation (3.1), parameter μ is fixed and that $c^{(0)} = (c_{10}, c_{20})$ is a solution of (3.4) and hence of (3.3). Let $c^{(1)} = (c_1, c_2)$ be a point in a neighborhood of $c^{(0)}$. Denote the two components of solution $x(t, \mu, c^{(1)})$ by $x_1(t, \mu, c_1, c_2)$ and $x_2(t, \mu, c_1, c_2)$. Since $M(t)$ has period T (both the columns of $M(t)$ are solutions of $\dot{x} = A(t)x$ which have period T), then if $\mu = 0$,

$$x_1(T, \mu, c_1, c_2) = c_1 \tag{5.2}$$

and

$$x_2(T, \mu, c_1, c_2) = c_2. \tag{5.3}$$

Since the terms in (3.1) have continuous second derivatives in μ, then by the Differentiability Theorem (1.11), solution $x(t, \mu, c)$ has a continuous second derivative in μ. Hence using (5.2) and (5.3) we have:

$$x_1(T, \mu, c_1, c_2) = c_1 + \mu g_1(c_1, c_2) + o(\mu), \tag{5.4}$$

$$x_2(T, \mu, c_1, c_2) = c_2 + \mu g_2(c_1, c_2) + o(\mu) \tag{5.5}$$

where $\lim_{\mu \to 0} o(\mu)/\mu = 0$. Let $c_1 = c_{10}$ and $c_2 = c_{20}$ in (5.4) and (5.5). By (3.3) we have:

$$c_{10} = x_1(T, \mu, c_{10}, c_{20}) = c_{10} + \mu g_1(c_{10}, c_{20}) + o(\mu), \tag{5.6}$$

$$c_{20} = x_2(T, \mu, c_{10}, c_{20}) = c_{20} + \mu g_2(c_{10}, c_{20}) + o(\mu). \tag{5.7}$$

Subtracting (5.6) and (5.7) from (5.4) and (5.5) we obtain:

$$x_1(T, \mu, c_1, c_2) - c_{10} = c_1 - c_{10} + \mu[g_1(c_1, c_2) - g_1(c_{10}, c_{20})] + o(\mu), \tag{5.8}$$

$$x_2(T, \mu, c_1, c_2) - c_{20} = c_2 - c_{20} + \mu[g_2(c_1, c_2) - g_2(c_{10}, c_{20})] + o(\mu). \tag{5.9}$$

Since the terms in (3.1) have continuous second derivatives in all variables, then g_1 and g_2 have continuous derivatives in c_1 and c_2. Let $\gamma_i = c_i - c_{i0}$ for $i = 1, 2$.

Then (5.8) and (5.9) become:

$$x_1(T, \mu, c_1, c_2) - c_{10} = \gamma_1 + \mu\gamma_1 g_{11} + \mu\gamma_2 g_{12} + \mu[o(\gamma_1, \gamma_2)] + o(\mu),$$

$$x_2(T, \mu, c_1, c_2) - c_{20} = \gamma_2 + \mu\gamma_1 g_{21} + \mu\gamma_2 g_{22} + \mu[o(\gamma_1, \gamma_2)] + o(\mu)$$

where $g_{ij} = \partial g_i/\partial c_j$ and $\lim_{\gamma \to 0} o(\gamma_1, \gamma_2) = 0$ where $\gamma = (\gamma_1^2 + \gamma_2^2)^{1/2}$. From the definition of characteristic exponents, the fact that $\mu o(\gamma_1, \gamma_2)$ and $o(\mu)$ are higher order terms and the Second Stability Theorem for a

Periodic Solution (4.18) it follows that if μ is sufficiently small and if the characteristic roots of:

$$\begin{pmatrix} 1 + \mu g_{11} & \mu g_{12} \\ \mu g_{21} & 1 + \mu g_{22} \end{pmatrix}$$

both have absolute value less than one then solution $x(t, \mu, c^{(0)})$ is asymptotically stable. Also from the Second Stability Theorem (4.18), we obtain: if the absolute values of the characteristic roots are both greater than one, then $x(t, \mu, c^{(0)})$ is unstable and if the number one separates the two absolute values, then $x(t, \mu, c^{(0)})$ is unstable. Thus we must study the roots z of the equation

$$\begin{vmatrix} 1 + \mu g_{11} - z & \mu g_{12} \\ \mu g_{21} & 1 + \mu g_{22} - z \end{vmatrix} = 0. \tag{5.10}$$

Let $(z - 1)/\mu = w$. Then (5.10) becomes:

$$\begin{vmatrix} \mu g_{11} - \mu w & \mu g_{12} \\ \mu g_{21} & \mu g_{22} - \mu w \end{vmatrix} = 0;$$

or dividing by μ^2, we have:

$$\begin{vmatrix} g_{11} - w & g_{12} \\ g_{21} & g_{22} - w \end{vmatrix} = 0. \tag{5.11}$$

To show the significance of (5.11), subtract c_1 from each side of (5.4) and c_2 from each side of (5.5) and obtain:

$$x_1(T, \mu, c_1, c_2) - c_1 = \mu g_1(c_1, c_2) + o(\mu), \tag{5.12}$$

$$x_2(T, \mu, c_1, c_2) - c_2 = \mu g_2(c_1, c_2) + o(\mu). \tag{5.13}$$

Taking $c_1 = c_{10}$, $c_2 = c_{20}$ in this pair of equations and using the fact that $x(T, \mu, c^{(0)}) = c^{(0)}$, we obtain:

$$0 = \mu g_1(c_{10}, c_{20}) + o(\mu), \tag{5.14}$$

$$0 = \mu g_2(c_{10}, c_{20}) + o(\mu). \tag{5.15}$$

Subtracting (5.14) and (5.15) from (5.12) and (5.13), respectively, we obtain:

$$\begin{aligned} x_1(T, \mu, c_1, c_2) - c_1 &= \mu[g_1(c_1, c_2) - g_1(c_{10}, c_{20})] + o(\mu), \\ x_2(T, \mu, c_1, c_2) - c_2 &= \mu[g_2(c_1, c_2) - g_2(c_{10}, c_{20})] + o(\mu) \end{aligned} \tag{5.16}$$

or

$$\begin{aligned} x_1(T, \mu, c_1, c_2) - c_1 &= \mu g_{11}\gamma_1 + \mu g_{12}\gamma_2 + o(\mu) + o(\gamma_1, \gamma_2), \\ x_2(T, \mu, c_1, c_2) - c_2 &= \mu g_{21}\gamma_1 + \mu g_{22}\gamma_2 + o(\mu) + o(\gamma_1, \gamma_2). \end{aligned} \tag{5.17}$$

Comparing (5.11), (5.16) and (5.17), we see that the Jacobian of $x(T, \mu, c) - c$ is, except for higher order terms in μ, γ_1, and γ_2, the determinant

$$J = \begin{vmatrix} \mu g_{11} & \mu g_{12} \\ \mu g_{21} & \mu g_{22} \end{vmatrix},$$

i.e., except for a factor μ^2 the determinant of the matrix

$$\begin{pmatrix} g_{11} & g_{12} \\ g_{21} & g_{22} \end{pmatrix},$$

and the characteristic roots of this matrix are the solutions w of (5.11). Now suppose the Jacobian of

$$x(T, \mu, c) - c$$

is positive at $c = c^{(0)}$. Then for all c sufficiently close to $c^{(0)}$, the Jacobian is positive. Assume first that the characteristic roots w_1, w_2, which are

TABLE I

$\frac{J(c^{(0)})}{\mu^2} > 0$	w_1, w_2 real	$w_1 > 0, w_2 > 0$ $w_1 > 0, w_2 > 0$ $w_1 < 0, w_2 < 0$ $w_1 < 0, w_2 < 0$	$\mu > 0$ $\mu < 0$ $\mu > 0$ $\mu < 0$	unstable asymp. stable asymp. stable unstable
$\frac{J(c^{(0)})}{\mu^2} > 0$	w_1, w_2 complex, i.e., $w_1 = a + ib$, $w_2 = a - ib$	$a > 0$ $a > 0$ $a < 0$ $a < 0$	$\mu > 0$ $\mu < 0$ $\mu > 0$ $\mu < 0$	unstable asymp. stable asymp. stable unstable
$\frac{J(c^{(0)})}{\mu^2} < 0$	Then w_1, w_2 are real and have opposite signs	z_1 and z_2 are real and $\lvert z_1\rvert$, $\lvert z_2\rvert$ are separated by 1.		unstable

solutions of (5.11), are real. Since the Jacobian is positive, then the product w_1, w_2 which is equal to J is positive. Hence w_1 and w_2 have the same sign. From the definition of w, we have:

$$z_i - 1 = \mu w_i \qquad (i = 1, 2),$$

$$(z_1 - 1)(z_2 - 1) = \mu^2 w_1 w_2$$

where z_1, z_2 are the solutions of (5.10). Since w_1, w_2 are real, then z_1, z_2 are real and if w_1, w_2 are both positive and if $\mu > 0$, then $z_1 > 1$ and $z_2 > 1$. Hence $x(t, \mu, c^{(0)})$ is unstable.

If w_1, w_2 are both negative, if $\mu > 0$ and if μ is sufficiently small, then $x(t, \mu, c^{(0)})$ is asymptotically stable. Applying this reasoning to the other cases, we may summarize the results obtained in the following table. (Some of these results hold only for μ sufficiently small.)

Now generally we will not be able to get the kind of detailed information about w_1 and w_2 described in the table above. We will be able to assume that the trace T of matrix

$$\begin{pmatrix} g_{11} & g_{12} \\ g_{21} & g_{22} \end{pmatrix},$$

which is equal to $w_1 + w_2$, is nonzero and we will be able to determine the sign of the trace. As Table I shows, these facts will yield the following information about the stability.

TABLE II

$\frac{J(c^{(0)})}{\mu^2} > 0$	$T > 0$	$\mu > 0$ $\mu < 0$	unstable asymp. stable
	$T < 0$	$\mu > 0$ $\mu < 0$	asymp. stable unstable
$\frac{J(c^{(0)})}{\mu^2} < 0$			unstable

With this information, we return to the study of equation (3.4) which by Theorem (3.8) becomes:

$$\mu \int_0^T [M(s)]^{-1}\{f[x(s, \mu, c), s, \mu]\}\, ds = 0. \tag{3.9}$$

Since we are dealing with the 2-dimensional case (i.e., an even dimensional case) then it follows from the Product Theorem (8.1) in Chapter I that the value of the local degree is not affected if the factor μ is disregarded. Let M_μ be the mapping:

$$M_\mu : c \rightarrow \int_0^T [M(s)]^{-1}\{f[x(s, \mu, c), s, \mu]\}\, ds.$$

Denote the fundamental matrix $M(t)$ by:

$$\begin{pmatrix} M_{11}(t) & M_{12}(t) \\ M_{21}(t) & M_{22}(t) \end{pmatrix},$$

and let the components of f be

$$f_1[x_1, x_2, t, \mu]$$

and

$$f_2[x_1, x_2, t, \mu]$$

where x_1, x_2 are the components of x. Dividing (3.9) by μ and then setting $\mu = 0$ in the integrand, we obtain:

$$\int_0^T \begin{pmatrix} M_{11}(s) & M_{12}(s) \\ M_{21}(s) & M_{22}(s) \end{pmatrix}^{-1} \begin{pmatrix} f_1[c_1 M_{11} + c_2 M_{12}, c_1 M_{21} + c_2 M_{22}, s, 0] \\ f_2[c_1 M_{11} + c_2 M_{12}, c_1 M_{21} + c_2 M_{22}, s, 0] \end{pmatrix} ds = 0. \tag{5.18}$$

In the remainder of the discussion, we use the following additional hypotheses:

(H-4) There is a solid circle S in the c_1c_2-plane such that $d[M_0, S, \bar{0}]$ is defined. (By the Invariance under Homotopy Theorem (6.4) of Chapter I, it follows that $d[M_\mu, S, \bar{0}]$ is defined and has the same value for all sufficiently small μ.)

(H-5) Functions f_1 and f_2 are polynomials in x_1 and x_2 and are continuous in μ.

(H-6) The trace of the matrix corresponding to the Jacobian of M_0 is not identically zero. (By (H-5), the trace is a polynomial in c_1 and c_2.)

(5.19) THEOREM. *Given $\varepsilon > 0$, then there is a continuous vector function*

$$k(t) = \begin{pmatrix} k_1(t) \\ k_2(t) \end{pmatrix}$$

such that $k_1(t)$ and $k_2(t)$ both have period T and $\max_t [|k_1(t)| + |k_2(t)|] < \varepsilon$ and such that if the term $f(x, t, \mu)$ in equation (3.1) is replaced by $f(x, t, \mu) + k(t)$, then at each solution (c_{10}, c_{20}) of the resulting equation (5.18), the Jacobian J and the trace T are both nonzero.

PROOF. First we apply Lemma (7.3a) of Chapter I.
Let

$$k(t) = \begin{pmatrix} aM_{11}(t) + bM_{12}(t) \\ aM_{21}(t) + bM_{22}(t) \end{pmatrix}.$$

Then a, b may be chosen so that

$$\max_t [|k_1(t)| + |k_2(t)|] < \varepsilon$$

and so that the solutions of the resulting equation (5.18) are all points at which $J \neq 0$. We want to insure also that $T \neq 0$ at each solution. We carry out the details of the proof for the case in which (5.18) has two solutions $e^{(1)} = (e_1^{(1)}, e_2^{(1)})$ and $e^{(2)} = (e_1^{(2)}, e_2^{(2)})$ at each of which $J \neq 0$. The argument goes through in the same way if there are n solutions.

Let N be a neighborhood of the origin $\bar{0}$. Since $J \neq 0$ at $e^{(1)}$ and $e^{(2)}$, there are disjoint neighborhoods N_1 of $e^{(1)}$ and N_2 of $e^{(2)}$ such that $J \neq 0$ on

$N_1 \cup N_2$ and such that $M_0(N_1 \cup N_2) \subset N$. By hypothesis (H-5), the trace is a polynomial in c_1 and c_2 which is not identically zero. Hence there is a point $\mathfrak{e}^{(1)} \in N_1$ such that $T \neq 0$ at $\mathfrak{e}^{(1)}$. Let $M_0(\mathfrak{e}^{(1)}) = (p_1, p_2) = p$. Since $J \neq 0$ on $N_1 \cup N_2$, then $M_0^{-1}(p)$ consists of the point $\mathfrak{e}^{(1)}$ and a point $\mathfrak{e}^{(2)}$ in N_2. Let

$$k(t) = \begin{pmatrix} \left(a + \frac{p_1}{T}\right)M_{11}(t) + \left(b + \frac{p_2}{T}\right)M_{12}(t) \\ \left(a + \frac{p_1}{T}\right)M_{21}(t) + \left(b + \frac{p_2}{T}\right)M_{22}(t) \end{pmatrix}.$$

The resulting equation (5.18) has two solutions $\mathfrak{e}^{(1)}$ and $\mathfrak{e}^{(2)}$ and $T \neq 0$ at $\mathfrak{e}^{(1)}$. Now repeat this procedure choosing the neighborhood N_1 of $\mathfrak{e}^{(1)}$ so that $J \neq 0$ on N_1 and $T \neq 0$ on N_1. ∎

Combining this last theorem and the In-the-large Implicit Function Theorem (10.2) of Chapter I, we have:

(5.20) THEOREM. *Let S be a solid circle such that $d[M_0, S, \bar{0}]$ is defined and nonzero. For definiteness, assume $d[M_0, S, \bar{0}] = d > 0$. Then there exist an $\varepsilon > 0$ and an $\varepsilon_1 > 0$ and a vector function $k(t)$ with components $k_1(t)$ and $k_2(t)$ such that:*

(1) *$k(t)$ is continuous and has period T;*
(2) $\max_t [|k_1(t)| + |k_2(t)|] < \varepsilon$;
(3) *if $f(x, t, \mu)$ in (3.1) is replaced by $f(x, t, \mu) + k(t)$, then at each solution of (5.18), the Jacobian J and the trace T are nonzero.*

If $f(x, t, \mu)$ in (3.1) is replaced by $f(x, t, \mu) + k(t)$, system (3.9) has a finite set of solutions, say m solutions, of the form

$$(c_1(\mu), c_2(\mu))$$

where $c_1(\mu)$, $c_2(\mu)$ are continuous functions of μ defined for μ such that $|\mu| \leqq \varepsilon_1$. Each point $(c_1(\mu), c_2(\mu))$ is contained in circle S and the Jacobian and the trace are both nonzero at each point. Also each pair of functions $c_1(\mu)$, $c_2(\mu)$ is distinct, i.e., if $(c_{11}(\mu), c_{12}(\mu))$ and $(c_{21}(\mu), c_{22}(\mu))$ are solutions of (3.9) and if μ_1, μ_2 are arbitrary points in $[-\varepsilon_1, \varepsilon_1]$, then

$$(c_{11}(\mu_1), c_{12}(\mu_1)) \neq (c_{21}(\mu_2), c_{22}(\mu_2)).$$

There are n solutions $(c_1(\mu), c_2(\mu))$ on which $J > 0$ and $n \geqq d$. Also there are $n - d$ solutions on which $J < 0$ and

$$n + (n - d) = m.$$

Combining this theorem with the information in Table II, we obtain:

(5.21) THEOREM. *If $d(M_0, S, \bar{0}) = d > 0$ and if $f(x, t, \mu)$ is replaced by $f(x, t, \mu) + k(t)$, then for each sufficiently small μ, equation (3.1) has $n(\geqq d)$*

distinct periodic solutions which are either asymptotically stable or unstable (depending on the sign of the trace) and $n - d$ solutions which are unstable. These periodic solutions are continuous functions of μ and if $x(t, \mu, c(\mu))$ is such a periodic solution, then

$$\lim_{\mu \to 0} x(t, \mu, c(\mu)) = M(t)c(0) + M(t) \int_0^t [M(s)]^{-1} g(s)\, ds.$$

In this discussion of stability, we have restricted ourselves to the 2-dimensional case. The same type of argument is valid in the higher-dimensional cases, but the conclusions are much more limited. (See Malkin [**1**, Chapter III].) The chief reason for this is that if the Jacobian is positive in the two-dimensional case, then we can conclude that the characteristic roots w_1, w_2 are both positive or both negative. In the n-dimensional case with $n \geqq 3$, there is a larger variety of possible conclusions, this variety becoming more extensive as n gets larger.

Now we consider the 2-dimensional resonance case in which $\dot{x} = A(t)x$ has just one linearly independent solution of period T. Let $c^{(0)} = (c_{10}, c_{20})$ be a solution of (3.4) for fixed μ. Let $c^{(1)} = (c_1, c_2)$ be near (c_{10}, c_{20}). Using the variation of constants formula (3.3a), we have:

$$\begin{aligned} (5.22) \qquad & x(T, \mu, c^{(1)}) - x(T, \mu, c^{(0)}) \\ &= M(T)c^{(1)} + \mu M(T) \int_0^T [M(s)]^{-1} f[x(s, \mu, c^{(1)}), s, \mu]\, ds \\ &\quad - M(T)c^{(0)} - \mu M(T) \int_0^T [M(s)]^{-1} f[x(s, \mu, c^{(0)}), s, \mu]\, ds \\ &= M(T) \begin{pmatrix} c_1 - c_{10} \\ c_2 - c_{20} \end{pmatrix} + \mu M(T) \int_0^T [M(s)]^{-1} \Big\{ f_x[x(s, \mu, c^{(0)}) \\ &\quad + \theta\{x(s, \mu, c^{(1)}) - x(s, \mu, c^{(0)})\}, s, \mu] \left[\theta \frac{\partial x}{\partial c}\right] \begin{pmatrix} c_1 - c_{10} \\ c_2 - c_{20} \end{pmatrix} \Big\}\, ds \end{aligned}$$

where $0 < \theta < 1$.

Let $c_1 - c_{10} = \gamma_1$, $c_2 - c_{20} = \gamma_2$. Since $\dot{x} = A(t)x$ has one solution of period T, we may assume that the first column of $M(t)$ has period T. Since $M(0)$ is the identity matrix, then

$$M(T) = \begin{pmatrix} 1 & b \\ 0 & d \end{pmatrix}$$

and $b \neq 0$ or $d \neq 1$. Matrix $M(T)$ can be computed because it depends only on the linear system $\dot{x} = A(t)x$.

Then (5.22) becomes

$$x(T, \mu, c^{(1)}) - x(T, \mu, c^{(0)})$$
$$(5.23) \qquad = \begin{pmatrix} 1 & b \\ 0 & d \end{pmatrix} \begin{pmatrix} \gamma_1 \\ \gamma_2 \end{pmatrix} + \mu \begin{pmatrix} g_{11} & g_{12} \\ g_{21} & g_{22} \end{pmatrix} \begin{pmatrix} \gamma_1 \\ \gamma_2 \end{pmatrix} + \mu o(\gamma_1, \gamma_2)$$

where

$$\lim_{(\gamma_1^2 + \gamma_2^2)^{1/2} \to 0} \frac{o(\gamma_1, \gamma_2)}{[\gamma_1^2 + \gamma_2^2]^{1/2}} = 0$$

and the g_{ij} are the appropriate derivatives. Thus the stability of $x(t, \mu, c^{(0)})$ is determined by the characteristic roots of the matrix,

$$\begin{pmatrix} 1 + \mu g_{11} & b + \mu g_{12} \\ \mu g_{21} & d + \mu g_{22} \end{pmatrix},$$

i.e., the roots z_1, z_2 of

$$(5.24) \qquad \begin{vmatrix} 1 + \mu g_{11} - z & b + \mu g_{12} \\ \mu g_{21} & d + \mu g_{22} - z \end{vmatrix} = 0.$$

Next we observe:

$$\begin{aligned} x(T, \mu, c^{(1)}) - c^{(1)} &= x(T, \mu, c^{(1)}) - x(0, \mu, c^{(1)}) \\ (5.25) \qquad &= [M(T) - M(0)]c^{(1)} \\ &\quad + \mu M(T) \int_0^T [M(s)]^{-1} f[x(s, \mu, c^{(1)}), s, \mu]\, ds. \end{aligned}$$

Take $c^{(1)} = c^{(0)}$ in this equation and we obtain:

$$\begin{aligned} x(T, \mu, c^{(0)}) - c^{(0)} &= 0 \\ (5.26) \qquad &= [M(T) - M(0)]c^{(0)} \\ &\quad + \mu M(T) \int_0^T [M(s)]^{-1} f[x(s, \mu, c^{(0)}), s, \mu]\, ds. \end{aligned}$$

Subtracting (5.26) from (5.25), we obtain:

$$\begin{aligned} &x(T, \mu, c^{(1)}) - x(0, \mu, c^{(1)}) \\ &\quad = [M(T) - M(0)](c^{(1)} - c^{(0)}) + \mu M(T) \int_0^T [M(s)]^{-1} \{ f_x[x(s, \mu, c^{(0)}) \\ &\qquad + \theta\{x(s, \mu, c^{(1)}) - x(s, \mu, c^{(0)})\}, s, \mu] \left[\theta \frac{\partial x}{\partial c}\right] (c^{(1)} - c^{(0)})\}\, ds \\ &\quad = \begin{pmatrix} 0 & b \\ 0 & d - 1 \end{pmatrix} \begin{pmatrix} \gamma_1 \\ \gamma_2 \end{pmatrix} + \mu \begin{pmatrix} g_{11} & g_{12} \\ g_{21} & g_{22} \end{pmatrix} \begin{pmatrix} \gamma_1 \\ \gamma_2 \end{pmatrix} + \mu o(\gamma_1, \gamma_2) \\ &\quad = \begin{pmatrix} \mu g_{11} & b + \mu g_{12} \\ \mu g_{21} & d - 1 + \mu g_{22} \end{pmatrix} \begin{pmatrix} \gamma_1 \\ \gamma_2 \end{pmatrix} + \mu o(\gamma_1, \gamma_2). \end{aligned}$$

Let w_1, w_2 denote the characteristic roots of the matrix

$$L = \begin{pmatrix} \mu g_{11} & b + \mu g_{12} \\ \mu g_{21} & d - 1 + \mu g_{22} \end{pmatrix}. \tag{5.27}$$

Comparing (5.24) and (5.27), we have:

$$z_1 = w_1 + 1,$$
$$z_2 = w_2 + 1.$$

As in the totally degenerate case, we can show that if a "small" vector function $k(t)$ is added to $f(x, t, \mu)$, then at each solution of (3.4), the determinant $|L|$ and the trace T of L are nonzero. Since the mapping whose local degree is computed is in this case, by the Reduction Theorem (10.1) of Chapter I, a mapping of a subset of R^1 into R^1, then the local degree is -1, 0, $+1$. The trace of L is

$$d - 1 + \mu(g_{11} + g_{22}).$$

If $d \neq 1$, then for μ sufficiently small, the sign of the trace T is the sign of $d - 1$. Thus we may obtain information analogous to that in Table II. We write out one case in detail: $d > 1$, i.e., $T > 0$.

TABLE III

$L > 0$	$w_1 > 0, w_2 > 0$	$z_1 > 1, z_2 > 1$	unstable
	$w_1 < 0, w_2 < 0$		asymp. stable or unstable
	$w_1 = a + ib$, $w_2 = a - ib$, $a > 0$	$\text{Re}(z_1) > 1$, $\text{Re}(z_2) > 1$	unstable
$L < 0$	w_1, w_2 are real and have opposite signs, say, $w_1 > 0$, $w_2 < 0$	$z_1 > 1, z_2 < 1$	unstable

From this, we may obtain a theorem analogous to Theorem (5.21).

Finally we remark that in the study of the existence and stability of periodic solutions of

$$\dot{x} = A(t)x + \mu f(x, t, \mu) + g(t) \tag{3.1}$$

we have often replaced $f(x, t, \mu)$ by $f(x, t, \mu) + k(t)$ where $k(t)$ has period T and is "arbitrarily small." That is, our conclusion described not the periodic solutions of (3.1) but those of

$$\dot{x} = A(t)x + \mu[f(x, t, \mu) + k(t)] + g(t).$$

From the point of view of the physicist, the introduction of $k(t)$ is not significant especially since $k(t)$ is part of the "forcing term," i.e., the periodic force impressed from the outside upon the system. The addition of $k(t)$ merely changes the description of this force arbitrarily slightly. Since any system of equations can be regarded only as an approximate description of the physical situation, the addition of the term $k(t)$ is a mere convenience. However, we cannot always take so cavalier an attitude. It is not always true that an arbitrarily small change in one of the terms of (3.1) does not make a significant change in the solutions of (3.1). For example, an arbitrarily small change in $A(t)$ may change our problem from a resonance to a nonresonance problem which may induce a radical change in the structure of the set of solutions. Also it may happen that $d[M_0, S, \bar{0}]$ is not defined. Later we will construct examples in which $d[M_0, S, \bar{0}]$ is not defined but an arbitrarily small change in $f(x, t, \mu)$ will result in $d[M_0, S, \bar{0}]$ being defined. But the value of $d[M_0, S, \bar{0}]$ will depend on how $f(x, t, \mu)$ is changed. Thus in what sense it is permissible to introduce the term $k(t)$ is really an open question. It is related to the theory of structural stability of autonomous systems. (See Lefschetz [**4**, pp. 239–245].)

6. **Some examples of quasilinear systems.** First some applications to subharmonic oscillations and entrainment of frequency in 2-dimensional systems will be described. The physical significance of these problems has been discussed by Minorsky [**1**] and Leimanis and Minorsky [**1**]. So only the mathematical aspect of the problems will be considered.

SUBHARMONIC OSCILLATIONS. We consider the system

$$\begin{aligned} \dot{x} &= y + \mu f_1(x, y, \sin mt, \cos mt, \mu) + g_1(t), \\ \dot{y} &= -x + \mu f_2(x, y, \sin mt, \cos mt, \mu) + g_2(t) \end{aligned} \tag{6.1}$$

where m is an integer such that $m \geqq 1$ and $g_1(t)$, $g_2(t)$ have period $2\pi/\nu$ where ν is an integer greater than one. Assume that f_1 and f_2 are polynomials in the indicated variables. In equation (6.1), $\dot{x} = A(t)x$ is

$$\begin{aligned} \dot{x} &= y, \\ \dot{y} &= -x \end{aligned}$$

and a fundamental matrix is

$$M(t) = \begin{pmatrix} \cos t & \sin t \\ -\sin t & \cos t \end{pmatrix}.$$

We want to determine if (6.1) has any solutions of (minimum) period 2π. Since the period of the "forcing term" consisting of $g_1(t)$ and $g_2(t)$ is $2\pi/\nu$, a number smaller than 2π, such a solution of period 2π is called a subharmonic oscillation. Following the theory developed in § 3, we remark first that $n - r = 0$, since $n = 2$, $r = 2$, i.e., this is a totally degenerate case. Hence we apply Theorem (3.8) and require that

$$\int_0^{2\pi} \begin{pmatrix} \cos t & -\sin t \\ \sin t & \cos t \end{pmatrix} \cdot \begin{pmatrix} g_1(t) \\ g_2(t) \end{pmatrix} dt = 0. \tag{6.2}$$

Note that this condition tends to permit the existence of subharmonic oscillations but to exclude harmonic oscillations (oscillations of period 2π in case $g_1(t)$ and $g_2(t)$ have period 2π). For example, the condition is obviously satisfied if

$$g_1(t) = \sin \nu t,$$
$$g_2(t) = 0$$

where ν is an integer greater than one. However, if

$$g_1(t) = \sin t,$$
$$g_2(t) = 0$$

then condition (6.2) is not satisfied.

Let $g_1(t) \equiv 0$ and $g_2(t) = \sin 2t$. For this case, equation (3.4) becomes:

$$\int_0^{2\pi} \begin{pmatrix} \cos s & -\sin s \\ \sin s & \cos s \end{pmatrix} \begin{Bmatrix} \mu f_1[x(s, \mu, c), s, \mu] \\ \mu f_2[x(s, \mu, c), s, \mu] \end{Bmatrix} ds = 0.$$

(We use here an abbreviated notation for the arguments of f_1 and f_2.) After the left side of the equation is multiplied by $1/\mu$, it defines a mapping M_μ of the (c_1c_2)-plane into itself. We compute $d[M_0, S, \bar{0}]$ and then apply Theorems (3.10) and (5.21). In order to study M_0, we must first obtain $x(s, 0, c)$, i.e., the solution of:

$$\dot{x} = y,$$
$$\dot{y} = -x + \sin 2t$$

which has the value $c = (c_1, c_2)$ at $t = 0$. By the variation of constants formula, this solution is:

$$x(t, 0, c) = c_1 \begin{pmatrix} \cos t \\ -\sin t \end{pmatrix} + c_2 \begin{pmatrix} \sin t \\ \cos t \end{pmatrix}$$
$$+ \begin{pmatrix} \cos t & \sin t \\ -\sin t & \cos t \end{pmatrix} \int_0^t \begin{pmatrix} \cos s & -\sin s \\ \sin s & \cos s \end{pmatrix} \cdot \begin{pmatrix} 0 \\ \sin 2s \end{pmatrix} ds$$

$$(6.3) \qquad = \begin{pmatrix} c_1 \cos t + c_2 \sin t \\ -c_1 \sin t + c_2 \cos t \end{pmatrix}$$
$$+ \begin{pmatrix} -\frac{2}{3} \sin^3 t \cos t - \frac{2}{3} \cos^3 t \sin t + \frac{2}{3} \sin t \\ -\frac{2}{3} \sin^4 t - \frac{2}{3} \cos^4 t + \frac{2}{3} \cos t \end{pmatrix}$$
$$= \begin{pmatrix} c_1 \cos t + c_2 \sin t \\ -c_1 \sin t + c_2 \cos t \end{pmatrix} + \begin{pmatrix} -\frac{1}{3} \sin 2t + \frac{2}{3} \sin t \\ \frac{2}{3} \cos 2t \end{pmatrix}.$$

The (minimum) period of $x(t, 0, c)$ is 2π. The solutions of (6.1) of period 2π "branch out" from $x(t, 0, c)$ as described in the last sentence of Theorem (5.21). Hence any solution of (6.1) of period 2π will have minimum period 2π. Thus any solutions obtained by applying Theorems (3.10) and (5.21) will describe subharmonic oscillations. The mapping

$$M_0 : (c_1, c_2) \to (c_1', c_2')$$

is described by

$$\begin{pmatrix} c_1' \\ c_2' \end{pmatrix} = \int_0^{2\pi} \begin{pmatrix} \cos s & -\sin s \\ \sin s & \cos s \end{pmatrix} \begin{pmatrix} f_1[x(s, 0, c), s, 0] \\ f_2[x(s, 0, c), s, 0] \end{pmatrix} ds.$$

Since f_1, f_2 are polynomials, then c_1' and c_2' are polynomials $P_1(c_1, c_2)$ and $P_2(c_1, c_2)$, respectively. In order to compute $d[M_0, S, \bar{0}]$ we first compute the terms of highest degree in c_1 and c_2. As shown in § 9 of Chapter I, if the sums of the terms of highest degree are homogeneous polynomials with no common real linear factors and if S is a circle with center at $\bar{0}$ and sufficiently large radius, then $d[M, S, \bar{0}]$ is defined and its value can be computed by studying the terms of highest degree.

Now suppose for $i = 1, 2$,

$$f_i(x, y, \sin mt, \cos mt, \mu) = h_i(x, y, \mu) + k_i(\sin mt, \cos mt, \mu) \tag{6.4}$$

where h_i and k_i are polynomials in the indicated variables. Then the nonconstant terms in $P_1(c_1, c_2)$ and $P_2(c_1, c_2)$ come from:

$$\int_0^{2\pi} \begin{pmatrix} \cos s & -\sin s \\ \sin s & \cos s \end{pmatrix} \begin{pmatrix} h_1[x(s, 0, c), 0] \\ h_2[x(s, 0, c), 0] \end{pmatrix} ds.$$

Writing out the components and using (6.3), we obtain:

$$\begin{aligned} P_{10} = \int_0^{2\pi} \{(\cos s)(h_1[c_1 \cos s + c_2 \sin s - \tfrac{1}{3} \sin 2s + \tfrac{2}{3} \sin s, \\ - c_1 \sin s + c_2 \cos s - \tfrac{2}{3} \cos 2s + \tfrac{2}{3} \cos s, 0]) \\ - (\sin s)(h_2[\qquad]\} \, ds, \end{aligned} \tag{6.5}$$

$$P_{20} = \int_0^{2\pi} \{(\sin s)(h_1[\qquad]) + (\cos s)(h_2[\qquad])\} \, ds.$$

If $g_1(t) \equiv g_2(t) \equiv 0$, then the components of $x(s, 0, c)$ are

$$c_1 \cos s + c_2 \sin s$$

and

$$-c_1 \sin s + c_2 \cos s.$$

Then P_{10} and P_{20} become:

$$P_{10} = \int_0^{2\pi} \{(\cos s)(h_1[c_1 \cos s + c_2 \sin s, -c_1 \sin s + c_2 \cos s]) \\ - (\sin s)(h_2[\quad])\}\, ds,$$

$$P_{20} = \int_0^{2\pi} \{(\sin s)(h_1[\quad]) + (\cos s)(h_2[\quad])\}\, ds.$$

Using the fact that

$$\int_0^{2\pi} (\sin s)^u (\cos s)^v \, ds \neq 0$$

if and only if u and v are both even, we see that in this case, polynomials P_{10} and P_{20} will contain only terms of odd exponent in c_1 and c_2 unless the coefficients in the nonconstant terms are all zero. Hence if P_{10} and P_{20} have no common real linear factors (i.e., if the local degree of mapping described by P_{10} and P_{20} is defined) then $d[M_0, S, \bar{0}]$ is odd and therefore nonzero. Hence by Theorem (3.10), there is at least one solution of period 2π for each sufficiently small μ. If $g(t)$ is not identically zero, e.g., if P_{10}, P_{20} are given by (6.5) then it is no longer true that P_{10}, P_{20} contain only terms in c_1, c_2 of odd exponent but the terms of highest exponent will be of odd exponent in c_1, c_2 as the following example indicates.

Let $h_1 = x^3$, $h_2 = y^3$. Then

$$\begin{aligned} P_{10} &= \int_0^{2\pi} \{(\cos s)[c_1 \cos s + c_2 \sin s - \tfrac{1}{3} \sin 2s + \tfrac{2}{3} \sin s]^3 \\ &\qquad - (\sin s)[-c_1 \sin s + c_2 \cos s - \tfrac{2}{3} \cos 2s + \tfrac{2}{3} \cos s]^3\}\, ds \\ &= \tfrac{3}{2}\pi c_1(c_1^2 + c_2^2) + \cdots \end{aligned}$$

where the dots represent terms in which the sum of the exponents of c_1 and c_2 is less than three.

$$\begin{aligned} P_{20} &= \int_0^{2\pi} \{(\sin s)[c_1 \cos s + c_2 \sin s - \tfrac{1}{3} \sin 2s + \tfrac{2}{3} \sin s]^3 \\ &\qquad + (\cos s)[-c_1 \sin s + c_2 \cos s - \tfrac{2}{3} \cos 2s + \tfrac{2}{3} \cos s]^3\}\, ds \\ &= \tfrac{3}{2}\pi c_2(c_1^2 + c_2^2) + \cdots \end{aligned}$$

where the dots have the same meaning as before.

The mapping $M : (c_1, c_2) \to (c_1'', c_2'')$ where

$$c_1'' = \tfrac{3}{2}\pi c_1(c_1^2 + c_2^2),$$
$$c_2'' = \tfrac{3}{2}\pi c_2(c_1^2 + c_2^2)$$

is such that $d[M, S, \bar{0}] = +1$, where S is any solid circle with center at $\bar{0}$.

Hence for sufficiently large S and sufficiently small μ,

$$d[M_\mu, S, \bar{0}] = +1.$$

Thus we can conclude that the system

$$\begin{aligned} \dot{x} &= y + \mu x^3, \\ \dot{y} &= -x + \mu y^3 + \sin 2t \end{aligned} \tag{6.6}$$

has at least one solution of period 2π for each sufficiently small μ. Note that in order to reach our conclusion, we have not used any properties of $g_1(t)$, $g_2(t)$ except that both have period π and that condition (6.2) is satisfied.

If condition (6.4) does not hold, then it is generally necessary to compute polynomials P_1 and P_2. As an example, let

$$\begin{aligned} f_1(x, y, \sin mt, \cos mt, \mu) &= (a_1x^2 + b_1xy + d_1y^2) \sin t + k_1(t), \\ f_2(x, y, \sin mt, \cos mt, \mu) &= (a_2x^2 + b_2xy + d_2y^2) \cos t + k_2(t) \end{aligned} \tag{6.7}$$

where k_1, k_2 have period 2π and let $g_1(t)$, $g_2(t)$ have period $2\pi/\nu$ where $\nu > 1$ and satisfy condition (6.2). We find the terms of highest exponent in c_1 and c_2 by computing the polynomials:

$$\begin{aligned} P_{11}(c_1, c_2) = \int_0^{2\pi} \{ & (\cos s \sin s)[a_1(c_1 \cos s + c_2 \sin s)^2 \\ & + b_1(c_1 \cos s + c_2 \sin s)(-c_1 \sin s + c_2 \cos s) \\ & + d_1(-c_1 \sin s + c_2 \cos s)^2] \\ & - (\sin s \cos s)[a_2(c_1 \cos s + c_2 \sin s)^2 \\ & + b_2(c_1 \cos s + c_2 \sin s)(-c_1 \sin s + c_2 \cos s) \\ & + d_2(-c_1 \sin s + c_2 \cos s)^2]\}\, ds \\ = \frac{\pi}{4} [& (b_2 - b_1)c_1^2 + [2(a_1 - a_2) - 2(d_1 - d_2)]c_1c_2 + (b_1 - b_2)c_2^2] \end{aligned}$$

$$\begin{aligned} P_{21}(c_1, c_2) = \int_0^{2\pi} \{ & (\sin^2 s)[a_1(c_1 \cos s + c_2 \sin s)^2 \\ & + b_1(c_1 \cos s + c_2 \sin s)(-c_1 \sin s + c_2 \cos s) \\ & + d_1(-c_1 \sin s + c_2 \cos s)^2] \\ & + (\cos s)^2[a_2(c_1 \cos s + c_2 \sin s)^2 \\ & + b_2(c_1 \cos s + c_2 \sin s)(-c_1 \sin s + c_2 \cos s) \\ & + d_2(-c_1 \sin s + c_2 \cos s)^2]\}\, ds \\ = \frac{\pi}{4} [& (a_1 + 3d_1 + 3a_2 + d_2)c_1^2 + (-2b_1 + 2b_2)c_1c_2 \\ & + (3a_1 + d_1 + a_2 + 3d_2)c_2^2]. \end{aligned}$$

Now we compute the local degree of the mapping described by $P_{11}(c_1, c_2)$ and $P_{21}(c_1, c_2)$. If $b_1 = b_2$, then

$$P_{11}(c_1, c_2) = \frac{\pi}{2}[(a_1 - a_2) - (d_1 - d_2)]c_1c_2,$$

$$P_{21}(c_1, c_2) = \frac{\pi}{4}[(a_1 + 3d_1 + 3a_2 + d_2)c_1^2 + (3a_1 + d_1 + a_2 + 3d_2)c_2^2].$$

Assume that

$$A = (a_1 - a_2) - (d_1 - d_2) \neq 0.$$

Let $B = a_1 + 3d_1 + 3a_2 + d_2$ and $C = 3a_1 + d_1 + a_2 + 3d_2$.

As shown in Chapter I, § 9, the mappings described by

$$\begin{aligned} c_1' &= c_1^2 - c_2^2 \\ c_2' &= 2c_1c_2 \end{aligned} \quad \text{and} \quad \begin{aligned} c_1' &= 2c_1c_2 \\ c_2' &= c_1^2 - c_2' \end{aligned}$$

have local degrees $+2$ and -2 at the origin and relative to any circle with center at the origin. If $A > 0$, $B > 0$, $C < 0$, then the homotopy

$$H_1(c_1, c_2t) = [A + t(1 - A)]c_1c_2,$$
$$H_2(c_1, c_2t) = [B + t(1 - B)]c_1^2 + [C + t(-1 - C)]c_2^2$$

shows that the mapping described by P_{11} and P_{21} has local degree -2. Similarly if $A < 0$, $B > 0$, $C < 0$, the local degree is $+2$. If B and C have the same sign, the local degree is 0.

If $b_1 \neq b_2$, we may easily compute the local degree in certain cases. If $a_1 = a_2$, $d_1 = d_2$, then

$$P_{11}(c_1, c_2) = \frac{\pi}{4}[(b_2 - b_1)c_1^2 + (b_1 - b_2)c_2^2],$$

$$P_{21}(c_1, c_2) = \frac{\pi}{4}[(4a_1 + 4d_1)c_1^2 + 2(b_2 - b_1)c_1c_2 + (4a_1 + 4d_1)c_2^2].$$

Setting $P_{21}(c_1,c_2) = 0$ and solving for c_1 in terms of c_2:

$$c_1 = \frac{-2(b_2 - b_1) \pm [4(b_2 - b_1)^2 - 4^3(a_1 + d_1)^2]^{1/2}}{2(4a_1 + 4d_1)} c_2. \tag{6.8}$$

Thus if $4^3(a_1 + d_1)^2 > 4(b_2 - b_1)^2$, then $P_{21}(c_1, c_2)$ has no real linear factors and the local degree is 0.

If $4^3(a_1 + d_1)^2 < 4(b_2 - b_1)^2$ and if the coefficients obtained in (6.8) are separated by $+1$ or -1 but not both $+1$ and -1, then since

$$P_{11}(c_1, c_2) = \frac{\pi}{4}(b_2 - b_1)[(c_1 - c_2)(c_1 + c_2)],$$

the local degree is $+2$ or -2.

If the local degree is 0, we can draw no conclusions. However, if the local degree is $+2$ or -2, we can conclude that the system:

$$\dot{x} = y + \mu[(a_1x^2 + b_1xy + d_1y^2)\sin t + k_1(t)] + g_1(t),$$
$$\dot{y} = -x + \mu[(a_2x^2 + b_2xy + d_2y^2)(\cos t) + k_2(t)] + g_2(t)$$

has at least one solution of period 2π for each sufficiently small μ. Note that this conclusion is obtained without imposing any conditions on $k_i(t)$, $g_i(t)$ for $i = 1, 2$ except that $g_1(t)$, $g_2(t)$ satisfy condition (6.2). By Theorem (5.21), if $k_1(t)$, $k_2(t)$ are varied "arbitrarily slightly," there exist at least two solutions of period 2π. If $g_1(t) \equiv g_2(t) \equiv 0$ and if $k_1(t) \equiv k_2(t) \equiv 0$, then we can easily study the stability of the solutions. Return to the case: $b_1 = b_2$. Assume $A > 0$, $B > 0$, $C < 0$. The local degree is -2. The Jacobian is

$$\frac{\pi^2}{8}\begin{vmatrix} Ac_2 & Ac_1 \\ 2Bc_1 & 2Cc_2 \end{vmatrix} < 0$$

for all (c_1, c_2) except the origin. Hence there are exactly two unstable solutions of period 2π for each sufficiently small μ if $k_1(t)$, $k_2(t)$ are varied "arbitrarily slightly." If $A < 0$, $B > 0$, $C < 0$, the local degree is $+2$ and the Jacobian is positive except at the origin. The trace is $Ac_2 + 2Cc_2$. Hence the solutions (c_1, c_2) which give rise to the periodic solutions are such that the trace has opposite signs at the two solutions. Hence there are exactly two periodic solutions, one asymptotically stable and one unstable, for each sufficiently small μ.

Entrainment of frequency. The problem is to investigate if there are oscillations in case the "forcing term" has period close to but not equal to the period of the solutions of the linear part of the system. We consider the system

$$\begin{aligned} \dot{x} &= y + \mu f_1(x, y, \sin \omega t, \cos \omega t, \mu), \\ \dot{y} &= -x + \mu f_2(x, y, \sin \omega t, \cos \omega t, \mu) \end{aligned} \tag{6.9}$$

where f_1, f_2 are polynomials in the indicated variables and ω is a function of μ such that

$$\omega = [1 + \mu\eta(\mu)]^{-1}$$

where $\eta(\mu)$ is a twice differentiable function of μ and $\eta(0) = \eta_0 \neq 0$. Let

$$\tau = \omega t = [1 + \mu\eta(\mu)]^{-1}t.$$

Then system (6.9) becomes:

$$\frac{dx}{d\tau} = (1 + \mu\eta(\mu))y + \mu(1 + \mu\eta(\mu))f_1(x, y, \sin \tau, \cos \tau, \mu),$$

$$\frac{dy}{d\tau} = -(1 + \mu\eta(\mu))x + \mu(1 + \mu\eta(\mu))f_2(x, y, \sin \tau, \cos \tau, \mu)$$

or

$$\frac{dx}{d\tau} = y + \mu[\eta(\mu)y + (1 + \mu\eta(\mu))f_1(x, y, \sin\tau, \cos\tau, \mu)],$$

$$\frac{dy}{d\tau} = -x + \mu[-\eta(\mu)x + (1 + \mu\eta(\mu))f_2(x, y, \sin\tau, \cos\tau, \mu)].$$

Assume that for $i = 1, 2$:

$$f_i(x, y, \sin\tau, \cos\tau, \mu) = h_i(x, y, \mu) + k_i(\sin\tau, \cos\tau, \mu) \tag{6.10}$$

where h_i and k_i are polynomials in the indicated variables. Then equation (3.4) becomes (dividing through by μ):

$$\int_0^{2\pi} \begin{pmatrix} \cos s & -\sin s \\ \sin s & \cos s \end{pmatrix} \begin{pmatrix} \eta(\mu)y + (1 + \mu\eta(\mu))[h_1 + k_1] \\ -\eta(\mu)x + (1 + \mu\eta(\mu))[h_2 + k_2] \end{pmatrix} ds = 0. \tag{6.11}$$

Let $\mu = 0$. As before, $x(s, 0, c)$ is the vector

$$\begin{pmatrix} c_1 \cos s + c_2 \sin s \\ -c_1 \sin s + c_2 \cos s \end{pmatrix}.$$

Then (6.11) becomes:

$$\begin{aligned} \int_0^{2\pi} \{(\cos s)[\eta_0(-c_1 \sin s + c_2 \cos s) & \\ + h_1(c_1 \cos s + c_2 \sin s, & \\ - c_1 \sin s + c_2 \cos s, 0) & \\ + k_1(\sin s, \cos s, 0)] & \\ - (\sin s)[-\eta_0(c_1 \cos s + c_2 \sin s) & \\ + h_2(c_1 \cos s + c_2 \sin s, & \\ - c_1 \sin s + c_2 \cos s, 0) & \\ + k_2(\sin s, \cos s, 0)]\}\, ds &= 0, \end{aligned}$$

$$\begin{aligned} \int_0^{2\pi} \{(\sin s)[\eta_0(-c_1 \sin s + c_2 \cos s) + h_1(\qquad) + k_1(\qquad)] & \\ + (\cos s)[-\eta_0(c_1 \cos s + c_2 \sin s) + h_2(\qquad) + k_1(\qquad)]\}\, ds &= 0 \end{aligned}$$

or

$$2\pi\eta_0 c_2 + H_1(c_1, c_2) + K_1 = 0,$$
$$-2\pi\eta_0 c_1 + H_2(c_1, c_2) + K_2 = 0$$

where H_i is a polynomial in c_1, c_2 such that each term of H_i contains one c_i with an odd exponent and the other c_i with an even exponent, and K_1, K_2 are constants.

The polynomials

$$H_1(c_1, c_2) + 2\pi\eta_0 c_2$$

and

$$H_2(c_1, c_2) - 2\pi\eta_0 c_1$$

are both of odd degree. Moreover, the local degree at the origin and relative to a sufficiently large circle with center at the origin of the mapping described by these polynomials is defined because even if all the terms in H_1 or H_2 vanish, the terms $2\pi\eta_0 c_2$ and $-2\pi\eta_0 c_1$ are nonzero. Since the polynomials are of odd degree, the local degree is odd and therefore nonzero. Hence we can conclude:

(6.12) THEOREM. *If condition* (6.10) *is satisfied, then for all sufficiently small* μ, *the system* (6.9) *has at least one solution of period*

$$2\pi/\omega = 2\pi/(1 + \mu\eta(\mu)).$$

This result is of some interest because we need make no assumption that the degree of the mapping is defined or is nonzero. As before, we need impose no hypotheses on the form of the functions $k_1(\sin \tau, \cos \tau, \mu)$ and $k_2(\sin \tau, \cos \tau, \mu)$.

In the examples thus far, we have considered only 2-dimensional systems. Analogous examples of n-dimensional systems ($n > 2$) in which the degree of degeneracy is two can also be obtained. For such examples, it is possible to study the solutions of the equations (3.7a) and (3.7b) directly, but a more effective method for the case in which the matrix A of equation (3.1) is a constant matrix is to assume that A is in canonical form. A particularly useful canonical form is described by Coddington and Levinson [**1**, Chapter 14] and has been used by Cronin [**5**] with the topological technique described here to obtain existence theorems for periodic solutions in the n-dimensional case ($n > 2$) if the degree of degeneracy is two.

We can obtain quite complete results in the 2-dimensional case (and in the n-dimensional case if the degree of degeneracy is two) because the mappings studied are mappings of the plane into itself and we have fairly complete techniques for computing the local degree of such mappings. However, higher dimensional cases can also be handled as the following example shows.

A THREE-DIMENSIONAL TOTALLY DEGENERATE SYSTEM. Consider the system

$$\dot{x}_1 = x_2 + \mu f_1(x_1, x_2, x_3, \cos t, \sin t, \mu),$$

$$\dot{x}_2 = -x_1 + \mu f_2(x_1, x_2, x_3, \cos t, \sin t, \mu),$$

$$\dot{x}_3 = \mu f_3(x_1, x_2, x_3, \cos t, \sin t, \mu).$$

The degree of degeneracy is three and a fundamental matrix is

$$M(t) = \begin{pmatrix} \cos t & \sin t & 0 \\ -\sin t & \cos t & 0 \\ 0 & 0 & 1 \end{pmatrix}.$$

As a simple example, let

$$f_1 = x_1^3 + k_1(\cos t, \sin t),$$
$$f_2 = x_2^3 + k_2(\cos t, \sin t),$$
$$f_3 = x_1^2 x_3 + k_3(\cos t, \sin t)$$

where k_1, k_2, k_3 are polynomials in the indicated variables. Since the branching equations (3.4) are:

$$\int_0^{2\pi} \{(\cos s)(c_1 \cos s + c_2 \sin s)^3 - (\sin s)(-c_1 \sin s + c_2 \cos s)^3\}\, ds = 0,$$

$$\int_0^{2\pi} \{(\sin s)(c_1 \cos s + c_2 \sin s)^3 + (\cos s)(-c_1 \sin s + c_2 \cos s)^3\}\, ds = 0,$$

$$\int_0^{2\pi} \{(c_1 \cos s + c_2 \sin s)^2\, c_3\}\, ds = 0;$$

then referring to example (6.6) in the discussion of subharmonic oscillations for the form of the first two branching equations, we see that we must compute the local degree of the mapping described by

$$P_{10}(c_1, c_2, c_3) = \tfrac{3}{2}\pi c_1(c_1^2 + c_2^2) + K_1,$$
$$P_{20}(c_1, c_2, c_3) = \tfrac{3}{2}\pi c_2(c_1^2 + c_2^2) + K_2,$$
$$P_{30}(c_1, c_2, c_3) = \pi(c_1^2 c_3 + c_2^2 c_3) + K_3$$

where K_1, K_2, K_3 are constants. Since P_{10}, P_{20}, P_{30} are all of odd degree, the local degree is odd and therefore nonzero. (Since the factor $(c_1^2 + c_2^2)$ is always positive except at the origin, it is easy to show the local degree is actually $+1$.)

An example in which the local degree is zero. Finally we consider a couple of examples which show the limitations of the local degree method. First we construct an example in which the local degree is zero. Our example shows that there may or may not be periodic solutions in this case. Consider the system:

$$\dot{x} = y + \mu y^2 \cos \omega t,$$
$$\dot{y} = -x + \mu[2x^2 \cos \omega t + \gamma \cos \omega t]$$

where $\omega = [1 + \mu\eta(\mu)]^{-1}$ as in example (6.9) and where γ is a constant. The mapping to be studied is described by:

$$\frac{\pi}{4}[c_1^2 + 3c_2^2 - 4c_1c_2 + 8\eta_0c_2],$$

$$\frac{\pi}{2}[3c_1^2 + c_2^2 - c_1c_2 - 4\eta_0c_1 + 2\gamma].$$

The local degree is 0 and if γ is sufficiently large, the branching equations have no real solutions if μ is sufficiently small. But if $\gamma = 0$, then since $\eta_0 \neq 0$, the local degree at 0 of the mapping relative to a sufficiently small circle with center at 0 is $+1$. Hence for sufficiently small μ, there is at least one periodic solution.

AN EXAMPLE IN WHICH THE LOCAL DEGREE IS NOT DEFINED. We consider:

$$\dot{x} = y + \mu f_1,$$
$$\dot{y} = -x + \mu f_2$$

in which the branching equations take the form:

$$P_1(c_1, c_2) + K_1 = 0,$$
$$P_2(c_1, c_2) + K_2 = 0$$

where P_1 and P_2 are homogeneous polynomials in c_1 and c_2 and K_1 and K_2 are constants. If the mapping of the (c_1, c_2)-plane into itself defined by these expressions is such that the local degree is not defined, the topological method is not applicable. That the local degree is not defined means that P_1 and P_2 have a common real linear factor. Hence if one of the coefficients of P_1 is varied, however slightly, the local degree will be defined. To vary P_1 or P_2 slightly, we need only vary f_1 or f_2 slightly. Thus if f_1 or f_2 is varied, however slightly, we obtain a system to which the topological method is applicable. The complication here is that the result obtained depends on how f_1 and f_2 are varied. It is easy to set up examples in which P_1 and P_2 have a common real linear factor and such that if f_1 is varied slightly one way the local degree of the mapping described by the new polynomials P_1 and P_2 is 0 and if f_1 is varied slightly in a different way the local degree of the mapping obtained is $+2$.

For example in the system (6.7) in the discussion of subharmonic oscillations, let $b_1 = b_2$, $a_1 = a_2 = 1$, $d_1 = 2$, and $d_2 = -2$. Then

$$a_1 + 3d_1 + 3a_2 + d_2 = 8,$$
$$3a_1 + d_1 + a_2 + 3d_2 = 0$$

and the mapping is described by

$$P_{11}(c_1, c_2) = -2\pi c_1c_2,$$
$$P_{21}(c_1, c_2) = 2\pi c_1^2$$

so that the local degree is not defined. If we change a_1 to .9, then

$$3a_1 + d_1 + a_2 + 3d_2 = 5.7 - 6 < 0$$

and the local degree is $+2$. If $a_1 = 1.1$, then

$$3a_1 + d_1 + a_2 + 3d_2 = 6.3 - 6 < 0$$

and the local degree is 0.

7. **Almost periodic solutions of quasilinear systems.** The local degree technique can also be used to investigate the existence of almost periodic solutions of quasilinear systems. The results obtained can be regarded as a qualitative solution of the problem proposed by Stoker [**1**, p. 239]. The study of almost periodic solutions is a classical problem which goes back to Poincaré and is considerably more difficult than the study of periodic solutions. However a systematic study of almost periodic solutions of quasilinear systems based on the work of Bogoliubov and Mitropolski [**1**] has been carried out by Malkin [**1**] (Hale [**1**] has also made significant extensions of the work of Bogoliubov and Mitropolski [**1**]) for systems of the form

$$\dot{x} = Ax + f(t) + \mu F(t, x, \mu) \tag{7.1}$$

where A is a constant matrix, functions f and F are almost periodic in t, and μ is a small parameter. The problem splits into two cases: that in which A has no criticial roots (i.e., no characteristic roots which are pure imaginary) and that in which A has critical roots. If A has no critical roots and if μ is sufficiently small, system (7.1) has a unique almost periodic solution. This is in analogy with the nonresonance case (Case I) of equation (3.1) in the study of periodic solutions. It was proved by Bogoliubov and Mitropolski [**1**].

The case in which A has critical roots is more complicated because there may be several almost periodic solutions or none. Malkin has reduced the problem to that of finding certain solutions of a system of m equations in m unknowns where m is the number of critical roots. These equations are analogous to but not the same as the system (3.4) of branching equations studied in the previous sections in order to find periodic solutions, and they are derived by different and lengthier considerations. Consequently we shall merely state the results from Malkin [**1**] as they are needed.

The topological technique we use is similar to that used in previous sections to study periodic solutions. The essential difference is that in order to prove the existence of periodic solutions, it is only necessary to show that a certain local degree is nonzero whereas to prove the existence of almost periodic solutions, we must first vary function F. Although the variations in F are "arbitrarily small," they are fairly elaborate. Function F regarded as a function of t must be varied and also regarded as a function of x. In physical language, the external force must be varied and the system itself must also be varied.

We first describe Malkin's result for (7.1). The hypotheses on the n-dimensional system (7.1) are:

(H-1) Matrix A has m critical roots where $m > 0$;

(H-2) the components of $f(t)$ are finite trigonometric sums;

(H-3) the components F_s of function F have the form:

$$F_s(t, x, \mu) = \sum_{i=0}^{\infty} F_s^{(i)}(t, x)\mu^i$$

where the series converges for positive μ such that $\mu < \mu_0$ where μ_0 is a fixed positive number;

(H-4) each $F_s^{(i)}$ has the form

$$F_s^{(i)} = A_{so}^{(i)} + \sum_p (A_{sp}^{(i)} \cos \nu_p t + B_{sp}^{(i)} \sin \nu_p t),$$

where the sum is finite and the coefficients are polynomials in the components of x.

We also assume that the number of sets of solutions of

$$\dot{x} = Ax \tag{7.2}$$

corresponding to each critical root of A is equal to the multiplicity of the critical root, i.e., that the Jordan canonical form of A is a diagonal matrix. Then (see Malkin [**1**, pp. 100–109, 282, 285]) equation (7.2) has m almost periodic solutions $\phi_1, \cdots, \phi_m$ and the system conjugate to (7.2) has m almost periodic solutions $\psi_1, \cdots, \psi_m$ and these solutions can be chosen so that $\phi_i \cdot \psi_j = \delta_{ij}$. Further we assume that:

$$\lim_{t\to\infty} \frac{1}{t}\int_0^t f(t)\cdot\psi_i(t)\, dt = 0 \qquad (i = 1, \cdots, m).$$

Then the generating system

$$\dot{x} = Ax + f(t) \tag{7.3}$$

has the almost periodic solution:

$$x^{(0)}(t) = \sum_{i=1}^{m} M_i\phi_i + x^{(0)*}(t)$$

where $M_1, \cdots, M_m$ are constants and $x^{(0)*}(t)$ is a particular almost periodic solution of (7.3).

(7.4) THEOREM (MALKIN'S THEOREM). *If the numbers* $M_1^{(0)}, \cdots, M_m^{(0)}$ *satisfy the equations*

$$P_i(M_1, \cdots, M_m) = \lim_{t\to\infty} \frac{1}{t}\int_0^t F(t, x^{(0)}, 0)\cdot\psi_i(t)\, dt = 0 \tag{7.5}$$
$$(i = 1, \cdots, m),$$

(which we call the branching equations) and if the characteristic roots of the matrix

$$\left(\frac{\partial P_i}{\partial x_j}\right),$$

evaluated at $(M_1^{(0)}, \cdots, M_m^{(0)})$, *all have nonzero real parts, then for sufficiently small* μ, *equation* (7.1) *has an almost periodic solution* $x(t, \mu)$ *such that*

$$\lim_{\mu \to 0} x(t, \mu) = \sum_{i=1}^{m} M_i^{(0)} \phi_i + x^{(0)*}(t).$$

The remainder of our discussion is a study of the branching equations (7.5). First from the hypotheses on (7.1), it follows that $P_1, \cdots, P_m$ are polynomials in $M_1, \cdots, M_m$. We must study the characteristic roots of

$$\left(\frac{\partial P_i}{\partial x_j}\right),$$

i.e., the roots $\lambda_1, \cdots, \lambda_m$ of

$$(7.6) \qquad f(\lambda) = \begin{vmatrix} \dfrac{\partial P_1}{\partial M_1} - \lambda & \dfrac{\partial P_1}{\partial M_2} & \cdots & \dfrac{\partial P_1}{\partial M_m} \\ \dfrac{\partial P_2}{\partial M_1} & \dfrac{\partial P_2}{\partial M_2} - \lambda & \cdots & \dfrac{\partial P_2}{\partial M_m} \\ & \cdots & & \\ \dfrac{\partial P_m}{\partial M_1} & \dfrac{\partial P_m}{\partial M_2} & \cdots & \dfrac{\partial P_m}{\partial M_m} - \lambda \end{vmatrix} = 0$$

where the $\partial P_i / \partial M_j$ are all evaluated at a solution of (7.5). Writing (7.6) in the form:

$$(7.7) \qquad f(\lambda) = \lambda^m + a_1 \lambda^{m-1} + \cdots + a_{m-1}\lambda + a_m = 0,$$

we have: each a_i $(i = 1, \cdots, m)$ is a polynomial in the $\partial P_i / \partial x_j$ $(i, j = 1, \cdots, m)$ and consequently each a_i is a polynomial in $M_1, \cdots, M_m$.

Now define $a_j = 0$ for $j > m$ and introduce the following determinants:

$$\delta_1 = a_1$$

$$\delta_2 = \begin{vmatrix} a_1 & a_3 \\ 1 & a_2 \end{vmatrix}$$

$$\cdots\cdots\cdots$$

$$\delta_k = \begin{vmatrix} a_1 & a_3 & \cdots & a_{2k-1} \\ 1 & a_2 & \cdots & a_{2k-2} \\ 0 & a_1 & \cdots & a_{2k-3} \\ \vdots & & & \vdots \\ 0 & & \cdots & a_k \end{vmatrix} \qquad (k = 3, \cdots, m).$$

(7.8) LEMMA. *Suppose that for a set of fixed values* $M_1, \cdots, M_m$,

$$\delta_k \neq 0 \quad (k = 1, 2, \cdots, m).$$

Then all the solutions of (7.7) *have nonzero real parts.*

PROOF. This is a simple corollary of a well-known theorem. (See Marden [**1**, p. 141, Theorem (40, 2)].)

(7.9) LEMMA. *If* $Q(y_1, \cdots, y_m)$ *is a polynomial in real variables* $y_1, \cdots, y_m$ *and* Q *is not identically zero; if* N_ε *is a neighborhood of radius* ε, *where* ε *is arbitrary, of a point* $(y_1^{(0)}, \cdots, y_m^{(0)})$ *in real Euclidean m-space; then there is a nonempty open set* $N \subset N_\varepsilon$ *such that if* $(\xi_1, \cdots, \xi_m) \in N$, *then*

$$Q(\xi_1, \cdots, \xi_m) \neq 0.$$

PROOF. Since Q is continuous, it is sufficient to show that there is a point $(x_1, \cdots, x_m) \in N_\varepsilon$ such that

$$Q(x_1, \cdots, x_m) \neq 0.$$

Suppose there is no such point $(x_1, \cdots, x_m)$. Then $\mathcal{Q}(y_1) = 0$ where

$$\mathcal{Q}(y_1) = Q(y_1, y_2^{(0)}, \cdots, y_m^{(0)})$$

for all points in N_ε of the form $(y_1, y_2^{(0)}, \cdots, y_m^{(0)})$ and hence polynomial $\mathcal{Q}(y_1)$ is identically zero. Each coefficient C in $\mathcal{Q}(y_1)$ is a polynomial in $y_2^{(0)}, \cdots, y_m^{(0)}$, i.e.,

$$C = C(y_2^{(0)}, \cdots, y_m^{(0)}).$$

From our assumption, we must have:

$$\mathscr{C}(y_2) = C(y_2, y_3^{(0)}, \cdots, y_m^{(0)}) = 0$$

for all $(y_2, y_3^{(0)}, \cdots, y_m^{(0)})$ which are coordinates of points in N_ε. Hence polynomial $\mathscr{C}(y_2)$ is identically zero. Repeating these arguments, we conclude that $Q(y_1, \cdots, y_m)$ is identically zero, contrary to hypothesis. ▮

DEFINITION. Let $\varepsilon > 0$. A real number r is said to be *varied less than* η if r is replaced by $r + \eta$ where η is a real number such that $\eta > 0$.

(7.10) LEMMA. *Let* $\varepsilon > 0$. *If the coefficients of the linear terms in* $P_1, \cdots, P_m$ *are varied less than* ε, *the resulting* $\delta_1, \cdots, \delta_m$, *regarded as polynomials in* $M_1, \cdots, M_m$ *are not identically zero.*

PROOF. For $i = 1, \cdots, m$,

$$P_i = K_i + b_{i1}M_1 + \cdots + b_{im}M_m + T_i^{(2)}$$

where $T_i^{(2)}$ consists of terms of order higher than one in $M_1, \cdots, M_m$ and K_i is a constant term. Then

$$\frac{\partial P_i}{\partial M_j} = b_{ij} + T_i^{(1)}$$

where $T_i^{(1)}$ consists of terms of order one or higher in $M_1, \cdots, M_m$. Thus the constant term of each δ_k, regarded as a polynomial in $M_1, \cdots, M_m$, is a polynomial in the numbers b_{ij} $(i, j = 1, \cdots, m)$. In fact this constant term is the corresponding determinant for the characteristic equation of the matrix (b_{ij}). Denote this corresponding determinant by Δ_k. Each Δ_k is a polynomial in the characteristic roots of (b_{ij}) and this polynomial is not identically zero. From the existence of the Jordan canonical form, it follows that each of these Δ_k's is therefore a polynomial in the b_{ij}'s which is not identically zero. Hence by Lemma (7.9) if the b_{ij} are varied less than ε, the constant term in δ_k (where δ_k is regarded as a polynomial in $M_1, \cdots, M_m$) is nonzero. Hence each polynomial δ_k is not identically zero. ∎

(7.11) Assumption. We assume all the solutions of the branching equations (7.5) are in the interior of a sphere S where

$$S = \left[(M_1, \cdots, M_m) \Big/ \sum_{i=1}^{m} M_i^2 \leqq R^2\right]. \tag{7.12}$$

(7.13) Lemma. *Let Assumption* (7.11) *be satisfied and let* $\varepsilon > 0$. *If the coefficients of the linear terms in* $P_1, \cdots, P_m$ *and the constant terms in* $P_1, \cdots, P_m$ *are varied less than* ε, *then if there exist solutions of the branching equations* (7.5) *they are isolated and at each solution the roots of* (7.9) *all have nonzero real parts.*

Proof. By Lemmas (7.8) and (7.10) we need only show that if the constant terms in $P_1, \cdots, P_m$ are varied less than ε, then each δ_k (regarded as a polynomial in $M_1, \cdots, M_m$) is nonzero at each of the solutions of the branching equations (7.5). As before, denote the constant terms in $P_1, \cdots, P_m$ by $K_1, \cdots, K_m$. From Lemma (7.3a) of Chapter I it follows that if $K_1, \cdots, K_m$ are varied less than ε, then the solutions of (7.5) are all points which are not critical points. Then the set of solutions of (7.5) is a finite set of points $(M_1^{(1)}, \cdots, M_m^{(1)}), \cdots, (M_1^{(p)}, \cdots, M_m^{(p)})$ at each of which the determinant $|\partial P_i / \partial M_j|$ is nonzero. Hence $\mathscr{M}$, the mapping described by the left sides in equations (7.5), is a homeomorphism in some neighborhood of each of the points $(M_1^{(1)}, \cdots, M_m^{(1)}), \cdots, (M_1^{(p)}, \cdots, M_m^{(p)})$. Thus there are p pairwise disjoint open sets $U_1, \cdots, U_p$ such that

$$(M_1^{(j)}, \cdots, M_m^{(j)}) \in U_j \qquad (j = 1, \cdots, p)$$

and an open set U such that

$$\bar{0} \in U$$

(where $\bar{0}$ is the origin of the real Euclidean m-space), such that M is a homeomorphism from U_j onto U $(j = 1, \cdots, p)$ and such that U is contained in a neighborhood of $\bar{0}$ of radius ε. By Lemma (7.10) the expression δ_1, regarded as a polynomial in $M_1, \cdots, M_m$, is not identically zero. Hence by Lemma (7.9), there is a neighborhood $V_1 \subset U_1$ such that $\delta_1 \neq 0$ on V_1.

Since δ_2, regarded as a polynomial in $M_1, \cdots, M_m$, is not identically zero, then by Lemma (7.9), there is a neighborhood $V_2 \subset U_1$ such that $\delta_2 \neq 0$ on V_2. Continuing in this way, we obtain: there is a neighborhood V_m such that $V_m \subset U_1$ and such that $\delta_k \neq 0$ on V_m for $k = 1, \cdots, m$.

Now $\mathscr{M}(V_m) \subset U$ and $\mathscr{M}^{-1}[\mathscr{M}(V_m)]$ consists of p disjoint open sets $W_1(= V_m), W_2, \cdots, W_p$ such that

$$W_j \subset U_j \qquad (j = 1, \cdots, p)$$

and such that

$$\delta_k \neq 0$$

on W_1 for $k = 1, \cdots, m$. Carrying this procedure $(p - 1)$ more times, we obtain the conclusion of Lemma (7.13). ∎

From Theorem (7.4) (Malkin's Theorem) and Lemma (7.13) we obtain:

(7.14) THEOREM. *If all the solutions of the branching equations* (7.5) *are inside a sphere S described by* (7.12) *and if the constant terms and the coefficients of the linear terms of $P_1, \cdots, P_m$ are varied less than ε where ε is an arbitrarily given positive number then corresponding to each solution of* (7.5) *there is an almost periodic solution of* (7.1) *for sufficiently small μ (as described in the statement of Theorem* (7.4)).

(7.15) ASSUMPTION. Let P_{iq} be the homogeneous polynomial which is the sum of the terms of highest degree in polynomial P_i $(i = 1, \cdots, m)$. We assume: if $(M_1, \cdots, M_m) \neq \bar{0}$, then

$$\sum |P_{iq}|^2 > 0$$

at $(M_1, \cdots, M_m)$.

From Assumption (7.15) it follows that $d[\mathscr{M}, S, \bar{0}]$ exists, where $d[\mathscr{M}, S, \bar{0}]$ is the topological degree of $\mathscr{M}$ at $\bar{0}$ and relative to any sufficiently large sphere S of the form

$$S = \left[(M_1, \cdots, M_m) \Big/ \sum_{i=1}^{m} M_i^2 \leqq R^2\right].$$

Also it follows that all the solutions of (7.5) are inside a sphere S of sufficiently large radius.

By Lemma (7.13)

$$\left|\frac{\partial P_i}{\partial M_j}\right| \neq 0$$

at each solution of (7.5) after the variations described in the statement of Lemma (7.13) are carried out. Hence $d[\mathscr{M}, S, \bar{0}]$ is the sum of the signs of $|\partial P_i / \partial M_j|$ at the solutions of (7.5). This observation together with Theorem (7.14) yields:

(7.16) THEOREM. *If Assumption* (7.15) *is satisfied, if $d[\mathscr{M}, S, \bar{0}] \neq 0$, and if the constant terms and the coefficients in the linear terms of $P_1, \cdots, P_m$*

are varied less than ε, where ε is an arbitrarily given positive number, then the number of distinct almost periodic solutions of (7.1) *for sufficiently small μ is* $\geqq |d[\mathscr{M}, S, \bar{0}]|$.

If $m = 2$, then by applying the stability theorem of Malkin [**1**, p. 312] it follows that the sign of the topological degree $d[\mathscr{M}, S, \bar{0}]$ gives information about the stability of the almost periodic solutions. E.g., if $d[\mathscr{M}, S, \bar{0}] < 0$, there are at least $|d[\mathscr{M}, S, \bar{0}]|$ unstable solutions and if $d[\mathscr{M}, S, \bar{0}] > 0$, there are at least $d[\mathscr{M}, S, \bar{0}]$ solutions which are all stable or are all unstable.

It can be shown that changes in the coefficients in the linear terms of $P_1, \cdots, P_m$ and the constants in $P_1, \cdots, P_m$ can be induced by changes of the same magnitude in the components of the function $F(t, x, \mu)$ because of the hypotheses (H-3) and (H-4).

Some examples. Consider the single second-order equation

$$\ddot{x} + x = \mu F(t, x, \dot{x}, \mu).$$

We write this as a 2-dimensional first-order system:

$$\dot{x} = y, \qquad \dot{y} = -x + \mu F(t, x, y, \mu). \tag{7.17}$$

Equation $\dot{x} = Ax$ is then:

$$\dot{x} = y, \qquad \dot{y} = -x \tag{7.18}$$

and the general solution of (7.18) is:

$$M_1 \begin{pmatrix} \cos t \\ -\sin t \end{pmatrix} + M_2 \begin{pmatrix} \sin t \\ \cos t \end{pmatrix}$$

where M_1, M_2 are arbitrary constants. Substituting

$$x = M_1(t) \cos t + M_2(t) \sin t,$$
$$y = -M_1(t) \sin t + M_2(t) \cos t$$

into (7.17), we obtain:

$$(\cos t) \frac{dM_1}{dt} + (\sin t) \frac{dM_2}{dt} = 0,$$

$$(-\sin t) \frac{dM_1}{dt} + (\cos t) \frac{dM_2}{dt} = \mu F(t, M_1 \cos t + M_2 \sin t, -M_1 \sin t + M_2 \cos t, \mu).$$

Solving for dM_1/dt and dM_2/dt, we obtain:

$$\frac{dM_1}{dt} = \begin{vmatrix} 0 & \sin t \\ \mu F & \cos t \end{vmatrix} = -\mu(\sin t)\{F(t, M_1 \cos t + M_2 \sin t, \\ - M_1 \sin t + M_2 \cos t, \mu)\},$$

$$\frac{dM_2}{dt} = \begin{vmatrix} \cos t & 0 \\ -\sin t & \mu F \end{vmatrix} = \mu(\cos t)\{F(t, M_1 \cos t + M_2 \sin t, \\ - M_1 \sin t + M_2 \cos t, \mu)\}.$$

Then

$$P_1(M_1, M_2) = \lim_{t \to \infty} \frac{1}{t} \int_0^t [-(\sin t)F(s, M_1 \cos s + M_2 \sin s, \\ - M_1 \sin s + M_2 \cos s, \mu)]\, ds,$$

$$P_2(M_1, M_2) = \lim_{t \to \infty} \frac{1}{t} \int_0^t [(\cos s)F(s, M_1 \cos s + M_2 \sin s, \\ - M_1 \sin s + M_2 \cos s, \mu)]\, ds.$$

Now suppose

$$F(t, x, y, \mu) = P_1(x, y) + P_2(t)$$

where P_1 is a polynomial in x and y and $P_2(t)$ is a function of t only. From the fact that

$$\int_0^{2\pi} (\sin s)^m (\cos s)^n\, ds \neq 0$$

if and only if m and n are both even, it follows that

$$\lim_{t \to \infty} \frac{1}{t} \int_0^t (\sin s)^m (\cos s)^n\, ds \neq 0$$

if and only if m and n are both even. Hence the corresponding $P_1(M_1, M_2)$ and $P_2(M_1, M_2)$ are polynomials in M_1, M_2 which contain constant terms and terms which are of odd degree in M_1 and M_2 but no others.

We assume that $P_1(M_1, M_2)$, $P_2(M_1, M_2)$ are nonconstant and that Assumption (7.15) is satisfied. Then $d(\mathscr{M}, S, \bar{0})$, where $\mathscr{M}$ is the mapping defined by P_1 and P_2, is odd and therefore nonzero. Applying Theorem (7.16), we conclude that if the nonlinear terms in (7.17) are changed arbitrarily slightly (i.e., if $\mu g_1(t)$, $\mu g_2(t)$ are added to the first and second equations respectively of (7.17)) then for sufficiently small μ, system (7.17) has at least one almost periodic solution $\Phi(t, \mu)$ such that

$$\lim_{\mu \to 0} \Phi(t, \mu) = M_1 \begin{pmatrix} \cos t \\ -\sin t \end{pmatrix} + M_2 \begin{pmatrix} \sin t \\ \cos t \end{pmatrix}$$

where M_1, M_2 are constants.

Thus if (7.17) has the form

$$\dot{x} = y,$$
$$\dot{y} = -x + \mu[P_1(x, y) + P_2(t)]$$

where $P_1(x, y)$ contains terms of odd degree and if the resulting polynomials $P_1(M_1, M_2)$, $P_2(M_1, M_2)$ are nonconstant, then if the nonlinear terms in (7.17) are varied arbitrarily slightly, there is at least one almost periodic solution for each sufficiently small μ. On the other hand if $P_1(x, y)$ contains only terms of even degree, then $P_1(M_1, M_2)$ and $P_2(M_1, M_2)$ are either constants or identically zero. If $P_1(M_1, M_2)$ or $P_2(M_1, M_2)$ is a nonzero constant, then system (7.5) has no solutions. Since finding the right kind of solution of (7.5) is only a sufficient condition for the existence of almost periodic solutions of (7.17) (see Malkin [**1**, p. 296]) our method in this case gives no information about whether (7.17) has almost periodic solutions.

If $F(t, x, y, \mu)$ is not of the form $P_1(x, y) + P_2(t)$ then we cannot make general conclusions like those above. We are usually forced in this case actually to compute the local degree. Suppose, for example,

$$F(t, x, y, \mu) = a(\sin t)x^2 + b(\cos t)(xy) + e(\cos t)(y^2) + h(t)$$

where a, b, e are constants and $h(t)$ is a function of t only. The constant terms in the resulting $P_1(M_1, M_2)$ and $P_2(M_1, M_2)$ depend on $h(t)$. Omitting these constant terms, we obtain:

$$P_1(M_1, M_2) = \tfrac{1}{8}[(b - a)M_1^2 + (-b - 3a)M_2^2 - 2cM_1M_2],$$
$$P_2(M_1, M_2) = \tfrac{1}{8}[(-b - c)M_1^2 = 2aM_1M_2 + (b + 3c)M_2^2].$$

Let $a > 0$ and suppose $b = a$. Then

$$P_1(M_1, M_2) = \tfrac{1}{8}[-4aM_2^2 - 2cM_1M_2] = \tfrac{1}{8}M_2[-aM_2 - 2cM_1],$$
$$P_2(M_1, M_2) = \tfrac{1}{8}[(a + 3c)M_2^2 + 2aM_1M_2 - (a + c)M_1^2].$$

Setting $P_1 = 0$, we obtain:

$$M_2 = 0$$

and

$$M_2 = -\frac{2c}{a}M_1.$$

Setting $P_2 = 0$, we obtain:

$$M_2 = \left\{\frac{-2a \pm \sqrt{4a^2 + 4(a + c)(a + 3c)}}{2(a + 3c)}\right\} M_1.$$

Assume $c > 0$. Using the technique described in Chapter I, we see that the local degree is $+2$ if

$$\frac{-2a - \sqrt{4a^2 + 4(a + c)(a + 3c)}}{2(a + 3c)} > -\frac{c}{2a}$$

or

$$-4a^2 - 2a\sqrt{4a^2 + 4(a + c)(a + 3c)} > -2c(a + 3c)$$

$$+2a\sqrt{4a^2 + 4(a + c)(a + 3c)} < +2c(a + 3c) - 4a^2.$$

If a has a fixed positive value, say $a = 1$, then if c is a sufficiently large positive number, this inequality is satisfied.

8. **Periodic solutions of autonomous quasilinear systems.** So far we have dealt entirely with nonautonomous systems, i.e., systems in which the independent variable t appears explicitly. Now we consider the problem of finding periodic solutions of an autonomous quasilinear system,

$$\dot{x} = Ax + \mu f(x, \mu) \tag{8.1}$$

where A is a constant matrix and the components of f are defined for all values of x and have continuous second derivatives with respect to μ and the components of x. At first glance, this problem might seem to be a special case of (3.1), the first equation we studied. There are two facts, however, that make the autonomous problem quite different from the nonautonomous problem.

First the very question of whether a periodic solution exists has to be altered. In the nonautonomous case, it was reasonable to look for a solution the period of which would be equal to the period of the functions f and g in (3.1). Functions f and g contain the "forcing terms," i.e., the terms which describe the forces from the outside which are imposed on the physical system. However, in the autonomous case, any periodic solution which occurs describes an oscillation in the physical system which arises "naturally" within the system itself. There is no reason to believe that as parameter μ is varied (and thus the system changed) the period of the solution will remain constant and, indeed, it does not remain constant. Thus in seeking periodic solutions, we must take into account the varying periods of the solutions. Secondly, we use the fact (which is the basis for the widely used phase plane method in the two-dimensional autonomous case) that if $x(t)$ is a solution of (8.1) which is defined for all t, then $x(t + \delta)$, where δ is an arbitrary real constant, is a solution of (8.1). To prove this, let $\tau = t + \delta$ and use the chain rule for differentiation as follows:

$$\frac{d}{dt}x(t + \delta) = \frac{d}{dt}x(\tau) = \frac{d}{d\tau}x(\tau)\frac{d\tau}{dt} = \frac{d}{d\tau}x(\tau)$$

$$= Ax(\tau) + \mu f[x(\tau), \mu] = Ax(t + \delta) + \mu f[x(t + \delta), \mu].$$

(This argument clearly breaks down if the system is nonautonomous.)

It turns out that when these facts are taken into consideration the treatment of the resonance case ($\dot{x} = Ax$ has nonzero periodic solutions) is quite different from the treatment of the resonance case for a nonautonomous equation. The branching equations obtained have a different form and we cannot apply directly our topological technique. Although we do not get as nice a general theory as in the nonautonomous case, we can obtain some new results. We describe these results for the three-dimensional case:

$$\dot{x} = Ax + \mu f(x, \mu) \tag{8.1}$$

where

$$A = \begin{pmatrix} 0 & 1 & 0 \\ 1 & 0 & 0 \\ 0 & 0 & 0 \end{pmatrix}. \tag{8.2}$$

Let $x(t, \mu, c)$ denote the general solution of (8.1). By the variation of constants formula,

$$x(t, \mu, c) = e^{tA}c + \mu \int_0^t e^{(t-s)A} f[x(s, \mu, c), \mu]\, ds.$$

As in the nonautonomous case we use this equation to write the condition of periodicity. From (8.2), it is clear that $\dot{x} = Ax$ has a solution of period 2π. Therefore, we seek a solution of period $2\pi + \tau$. As before (Theorem (3.2)), a necessary and sufficient condition that $x(t, \mu, c)$ have period $2\pi + \tau$ is that

$$(e^{(2\pi+\tau)A} - E)c + \mu \int_0^{2\pi+\tau} e^{(2\pi+\tau-s)A} f[x(s, \mu, c), \mu]\, ds = 0$$

or

$$\begin{aligned} &(e^{2\pi A} - E)c + e^{2\pi A}(e^{\tau A} - E)c \\ &\qquad + \mu \int_0^{2\pi+\tau} e^{(2\pi+\tau-s)A} f[x(s, \mu, c), \mu]\, ds = 0. \end{aligned} \tag{8.3}$$

We will presently see that (8.3) is a more convenient form. Since A has the form (8.2) then

$$e^{tA}c = \begin{pmatrix} \cos t & \sin t & 0 \\ -\sin t & \cos t & 0 \\ 0 & 0 & 1 \end{pmatrix} c.$$

Now if

$$c = \begin{pmatrix} c_1 \\ c_2 \\ c_3 \end{pmatrix},$$

the first component of $e^{tA}c$ is:

$$c_1 \cos t + c_2 \sin t.$$

If c_1, c_2 have the fixed values c_1^0, c_2^0 then there is a $t_0 \in [0, 2\pi]$ such that

$$c_1 \cos(t_0) + c_2 \sin(t_0) = 0.$$

Hence, for sufficiently small μ, the first component of solution $x(t, \mu, c^{(0)})$, where the vector $c^{(0)}$ has as its first two components c_1^0 and c_2^0, is zero for some t_1 near t_0. This follows from the continuity by the Existence Theorem for a System with a Parameter (1.8). The value t_1 depends on μ. Then since (8.1) is an autonomous function, $x(t - t_1, \mu, c^{(0)})$ is a solution of (8.1), and the first component of solution $x(t - t_1, \mu, c^{(0)})$ is zero at $t = 0$. Thus the problem becomes that of finding a solution the first component of which is zero at $t = 0$ and which has period $2\pi + \tau$.

Returning to equation (8.3), we note that $e^{2\pi A} = E$. Dividing (8.3) by μ, we have:

$$(8.4) \qquad e^{2\pi A}\left(\frac{e^{\tau A} - E}{\tau}\right) c \left(\frac{\tau}{\mu}\right) + \int_0^{2\pi+\tau} e^{(2\pi+\tau-s)A} f[x(s, \mu, c), \mu]\, ds = 0.$$

The limit as μ goes to zero of the first term on the left exists because the whole expression has the value zero and the limit as μ goes to zero of the second term clearly exists. Let μ go to zero. Then (8.4) becomes:

$$(8.5) \qquad e^{2\pi A} A c \lim_{\mu \to 0} \frac{\tau}{\mu} + \int_0^{2\pi} e^{(2\pi - s)A} f[e^{sA}c, 0]\, ds = 0.$$

Let $\nu = \lim_{\mu \to 0} \tau/\mu$ and note that $e^{2\pi A} = E$. Then (8.5) becomes:

$$\nu A c + \int_0^{2\pi} e^{(2\pi - s)A} f[e^{sA}c, 0]\, ds = 0.$$

Write the vector c as:

$$\begin{pmatrix} 0 \\ c_2 \\ c_3 \end{pmatrix}$$

and then (8.5) becomes:

$$\begin{pmatrix} 0 & 1 & 0 \\ -1 & 0 & 0 \\ 0 & 0 & 0 \end{pmatrix} \begin{pmatrix} 0 \\ c_2 \\ c_3 \end{pmatrix} \nu + \int_0^{2\pi} \begin{pmatrix} \cos s & -\sin s & 0 \\ \sin s & \cos s & 0 \\ 0 & 0 & 1 \end{pmatrix} \begin{pmatrix} f_1 \\ f_2 \\ f_3 \end{pmatrix} ds = 0.$$

Writing out the components of this vector equation, we obtain:

$$P_1(c_2, c_3, \nu) = \nu c_2 + \int_0^{2\pi} [(\cos s)(f_1) - (\sin s)(f_2)]\, ds = 0,$$

$$P_2(c_2, c_3,) = \int_0^{2\pi} [(\sin s)(f_1) + (\cos s)(f_2)]\, ds = 0,$$

$$P_3(c_2, c_3) = \int_0^{2\pi} [f_3]\, ds = 0$$

where $f_i = f_i[c_2 x^{(2)} + c_3 x^{(3)}, 0]$ and

$$x^{(2)} = \begin{pmatrix} -\sin t \\ \cos t \\ 0 \end{pmatrix} \text{ and } x^{(3)} = \begin{pmatrix} 0 \\ 0 \\ 1 \end{pmatrix}.$$

That is,

$$f_i = f_i(-c_2 \sin t, c_2 \cos t, c_3, 0).$$

Suppose f is a polynomial in all its variables. Let $\mathfrak{M}$ be the mapping defined by:

$$c_2' = P_2(c_2, c_3),$$
$$c_3' = P_3(c_2, c_3).$$

If $d[\mathfrak{M}, C, \bar{0}] \neq 0$, then the pair of equations

$$P_2(c_2, c_3) = 0,$$
$$P_3(c_2, c_3) = 0$$

has solutions even if μ is varied away from zero. If the component c_2 of a solution is different from zero, then the equation

$$P_1(c_2, c_3, \nu) = 0$$

can be solved for ν in terms of c_2 and c_3.

As an example, let x_1, x_2, x_3 be the components of the vector x and suppose

$$f_1 = -x_1 x_3,$$
$$f_2 = x_1 x_2,$$
$$f_3 = x_2^2 - x_3^2 + k$$

where k is a constant. Then

$$P_2 = \pi c_2 c_3,$$
$$P_3 = \pi(c_2^2 - 2c_3^2 + 2k).$$

Thus if C is any circle in the (c_2, c_3)-plane with center at $\bar{0}$ and sufficiently large radius, then $d[\mathfrak{M}, C, \bar{0}] = -2$. So there is at least one limit cycle if the component c_2 of the solution is nonzero so that the equation

$$P_1(c_2, c_3, \nu) = 0$$

can be solved.

It should be noticed that unlike the nonautonomous case, no information is obtained about the number of limit cycles or the stability properties of the limit cycles. (It might be possible to obtain an estimate on the number of limit cycles if we assume that the f_i's are functions of μ and then vary μ.)

9. **Periodic solutions of systems with a "large" nonlinearity.** So far we have dealt only with quasilinear systems. Now we will apply the methods of Chapter I to study the periodic solutions of systems with a "large" nonlinearity (for example the equation (3.1) with parameter μ set equal to one). The approach to the problem which will be used is due to Gomory [**1**]. It makes possible new results and the unification of known results. (In Chapter IV, the method of Cesari, which uses infinite-dimensional topological techniques, will be used to study periodic solutions of equations with large nonlinearities.)

As might be expected, this problem is far more difficult than the problem for quasilinear systems: we do not obtain as complete results even for the existence problem and we obtain no information about stability. The results are all for nonautonomous systems and form in part an analog of the Poincaré-Bendixson Theorem for autonomous systems. To make this relationship clear, we begin by describing the Poincaré-Bendixson Theorem. Then we obtain a topological analog for nonautonomous systems and describe the application of this topological result.

In order to state the Poincaré-Bendixson Theorem, we need a couple of preliminary definitions. Suppose that in the two-dimensional autonomous system

$$\begin{aligned} \dot{x} &= P(x, y), \\ \dot{y} &= Q(x, y) \end{aligned} \tag{9.1}$$

the functions P and Q have continuous first derivatives in a domain D in the x, y-plane. (This hypothesis insures that the Basic Existence Theorem (1.3) can be applied in D.)

DEFINITION. Let $(x(t), y(t))$ be a solution of (9.1) such that not both $x(t)$ and $y(t)$ are constants. The curve

$$\begin{aligned} x &= x(t), \\ y &= y(t) \end{aligned}$$

is called an *orbit* of (9.1).

It follows from the uniqueness condition of the Basic Existence Theorem (1.3) that through each point of domain D, there passes at most one orbit of (9.1). It should be observed that there is *not* a 1-1 correspondence between orbits and solutions of (9.1). For if an orbit is described by a solution $(x(t), y(t))$ of (9.1), then the orbit is also described by:

$$\begin{aligned} x &= x(t + \delta), \\ y &= y(t + \delta) \end{aligned}$$

where δ is an arbitrary real constant. But as shown in § 8 on autonomous quasilinear systems, the pair $(x(t + \delta), y(t + \delta))$ is also a solution of (9.1).

From the uniqueness condition of the Basic Existence Theorem (1.3), it follows that an orbit cannot cross itself. If there exist two values t_1, t_2 such that $t_1 \neq t_2$ and if for solution $(x(t), y(t))$, it is true that

$$x(t_1) = x(t_2),$$
$$y(t_1) = y(t_2)$$

then the orbit described by $(x(t), y(t))$ is a simple closed curve. The orbit is said to be *periodic*. Plainly each solution of (9.1) which describes a periodic orbit is a periodic solution of (9.1). The Poincaré-Bendixson Theorem describes conditions under which there exists a periodic orbit.

DEFINITION. A *positive semi-orbit* of (9.1) is a curve described by

$$x = x(t),$$
$$y = y(t)$$

where $t \geqq t_0$, a fixed real value, and where $(x(t), y(t))$ is a solution of (9.1).

From the Continuation Theorem (1.7), we have

(9.2) THEOREM. *Let D_1 be a bounded domain such that $\bar{D}_1 = D_1 \cup D_1'$ is contained in domain D. If C^+ is the point set corresponding to a positive semi-orbit of* (9.1) *and $C^+ \subset D_1 \cup D_1'$, and if $(x(t), y(t))$ is a solution of* (9.1) *which describes C^+, the solution $(x(t), y(t))$ is defined for all $t \geqq t_0$, a fixed real value.*

DEFINITION. If system (9.1) has a solution of the form

$$x(t) \equiv x_0,$$
$$y(t) \equiv y_0$$

where x_0, y_0 are constants, then the point (x_0, y_0) is a *critical point* of (9.1).

It is easy to see that the point (x_0, y_0) is a critical point if and only if

$$P(x_0, y_0) = Q(x_0, y_0) = 0.$$

DEFINITION. If the positive semi-orbit C^+ is described by a solution $(x(t), y(t))$ of (9.1) which is defined for all $t \geqq t_0$, a fixed real value, then the point (x_1, y_1) is a *limit point* of C^+ if there is a sequence $\{t_n\}$ such that $\lim_{n\to\infty} t_n = \infty$ and such that

$$\lim_{n\to\infty} (x(t_n), y(t_n)) = (x_1, y_1).$$

Let $L(C^+)$ denote the set of limit points of C^+.

(9.3) POINCARÉ-BENDIXSON THEOREM. *Suppose the positive semi-orbit C^+ of* (9.1) *is contained in $D_1 \cup D_1'$ where D_1 is a bounded domain and $D_1 \cup D_1' \subset D$. If $L(C^+)$ contains no critical points, then either*:

(i) *C^+ is a periodic orbit (and thus $C^+ = L(C^+)$), or*

(ii) *$L(C^+)$ is a periodic orbit.*

Intuitively, the Poincaré-Bendixson Theorem is reasonable. Arguing crudely, we may say: if there is no critical point for the orbit to approach, then since the orbit stays in a bounded set and is therefore described by a solution of (9.1) which is defined for all $t \geqq t_0$ (by Theorem (9.2)), then the orbit must either "double back" on itself and be periodic or the orbit must spiral toward a periodic orbit. Whether or not this argument seems reasonable the fact is that a formal proof of the Poincaré-Bendixson Theorem is a lengthy procedure. The proof can be found in most standard texts on ordinary differential equations. (See e.g., Coddington and Levinson [**1**, p. 389 ff.].)

Note that the Poincaré-Bendixson Theorem is true only in the 2-dimensional case. Exactly where the condition $n = 2$ is required in the proof is pointed out by Hurewicz [**1**, pp. 107, 109].

We proceed to the analogous study for a nonautonomous system and study the two-dimensional system

$$\begin{aligned} \dot{x} &= P(x, y) + E_1(t), \\ \dot{y} &= Q(x, y) + E_2(t) \end{aligned} \tag{9.4}$$

where $E_1(t)$, $E_2(t)$ have continuous first derivatives for all real t and both have period T. At least one of the functions $E_1(t)$, $E_2(t)$ is not identically zero, i.e., (9.4) is nonautonomous. Assume also that $P(x, y)$, $Q(x, y)$ have continuous first derivatives in x and y for all real x and y. A solution of (9.4) gives rise to a mapping of the xy-plane into itself in this way: let

$$u(t, u^0) = \{x(t, u^0), y(t, u^0)\}$$

be a solution of (9.4) which has the value u^0 at $t = 0$. For each value of t for which $u(t, u^0)$ exists, we may define a mapping f_t of the xy-plane into itself by:

$$f_t(u^0) = u(t, u^0).$$

We define a vector field in the xy-plane: let $t_0 > 0$ be a fixed value of t and assign to the point u^0 the vector

$$f_{t_0}(u^0) - u^0 = u(t_0, u^0) - u^0.$$

By letting t_0 vary we obtain a continuously varying family of vector fields which we denote by $V(t)$.

(9.5) Hypothesis. There is a 2-cell σ in the xy-plane such that the critical points of

$$\begin{aligned} \dot{x} &= P(x, y) + E_1(0), \\ \dot{y} &= Q(x, y) + E_2(0) \end{aligned} \tag{9.5a}$$

are all in the interior of σ and such that $V(t)$ is defined on σ for $t \in [0, T]$ and the index of $V(t)$ relative to the boundary σ' of σ is defined for all $t \in [0, T]$. That is, $V(t)$ has no critical points on the boundary of σ.

This hypothesis marks an essential difference between the quasilinear problem and the problem of large nonlinearity. In the quasilinear case, we had only to show that the local degree of the mapping defined by the branching equations at $\mu = 0$ is defined. Then from the Invariance under Homotopy Theorem (6.4) of Chapter I, it follows that the local degree is defined and has the same value for all sufficiently small μ. Now that we are considering the case of a large nonlinearity we must *assume* that the local degree (or index of the corresponding vector field) is defined over a large interval $[0, T]$. The hardest part of this topological approach is to prove that Hypothesis (9.5) is satisfied.

We will establish some existence theorems by using Hypothesis (9.5) and then return to the study of Hypothesis (9.5) itself. First we obtain a method for computing the index of $V(t)$.

(9.6) THEOREM. *If Hypothesis* (9.5) *is satisfied then for all* $t \in [0, T]$ *the index of* $V(t)$ *with respect to* σ' *equals the index with respect to* σ' *of the vector field* $V_1(x, y)$ *defined by*:

$$V_1(x, y) = (P(x, y) + E_1(0), Q(x, y) + E_2(0)).$$

PROOF. First by Hypothesis (9.5), the index of $V_1(x, y)$ relative to σ' is defined. Let $u^0 = (x^0, y^0)$ be a point on σ'. The components of $u(t, u^0)$ may be written:

$$x(t, u^0) = x(t, x^0, y^0) = x(0, x^0, y^0) + t\dot{x}(0, x^0, y^0) + \tfrac{1}{2}t^2\ddot{x}(t', x^0, y^0),$$

$$y(t, u^0) = y(t, x^0, y^0) = y(0, x^0, y^0) + t\dot{y}(0, x^0, y^0) + \tfrac{1}{2}t^2\ddot{y}(t'', x^0, y^0)$$

where $0 < t' < t$ and $0 < t'' < t$. Let $t_1 \leqq T$ be fixed. Then for $t < t_1$, $u(t, u^0)$ is in some bounded closed set R for all $u^0 \in \sigma'$ since $V(t)$ is defined and continuous on σ' by Hypothesis (9.5). But $\ddot{x}$ and $\ddot{y}$ are bounded on R since they are continuous on R. Hence

$$\lim_{t \to 0} \frac{u(t, u^0) - u^0}{t} = V_1(x^0, y^0)$$

and this limit is uniform on σ'. Hence for t sufficiently small, the angle between $V_1(x^0, y^0)$ and $V(t)$ is less than ε for all $(x^0, y^0) \in b(\sigma)$. So for t sufficiently small, the indices of $V_1(x^0, y^0)$ and $V(t)$ with respect to σ' are the same. From Hypothesis (9.5), the index of $V(t)$ with respect to σ' exists and has the same value for all $t \in [0, T]$. ∎

The index with respect to σ' of $V_1(x, y)$ is, by definition, the same as the local degree at $\bar{0}$ and relative to σ of the mapping:

$$M_1 : (x, y) \to (x', y')$$

defined by:

$$x' = P(x, y) + E_1(0),$$

$$y' = Q(x, y) + E_2(0).$$

(9.7) THEOREM *If Hypothesis* (9.5) *is satisfied and if* $d[M_1, \sigma, \bar{0}] \neq 0$, *system* (9.4) *has at least one solution of period* T.

PROOF. By Theorem (9.6), the index of $V(t)$ with respect to σ' is nonzero. In particular this holds for $V(T)$. Hence mapping f_T has a fixed point $u^{(1)} \in \sigma$. The solution of (9.4) through $u^{(1)}$ returns to $u^{(1)}$ after time interval T which is the period of $E_1(t)$ and $E_2(t)$. Hence the solution has period T. ∎

(9.8) THEOREM. *If Hypothesis* (9.5) *is satisfied and if* $d[M_1, \sigma, \bar{0}] = d \neq 0$, *then if* $E_1(t)$, $E_2(t)$ *are varied arbitrarily slightly* (*the exact meaning of "arbitrarily slight varying" is described in the proof*), *we obtain a system which has at least* $|d|$ *distinct solutions of period* T *in* σ *and has at most a finite number of distinct solutions of period* T *in* σ.

PROOF. By Theorem (9.6), the local degree at $\bar{0}$ and relative to σ of the mapping

$$f_T(u^0) - u^0 = u(T, u^0) - u^0$$

is d. We obtain more explicitly the form of this mapping. First $u(t, u^0)$ is the solution of (9.4) which has the initial value u^0 at $t = 0$. The corresponding system of integral equations is (letting $u^0 = (x^0, y^0)$):

$$x = x^0 + \int_0^t [P(x, y) + E_1(s)]\, ds,$$

$$y = y^0 + \int_0^t [Q(x, y) + E_2(s)]\, ds$$

and the successive approximations to the solution of the system are (see the proof of the Basic Existence Theorem (1.3)):

$$x^{m+1} = x^0 + \int_0^t \{P[x^m(s), y^m(s)] + E_1(s)\}\, ds,$$

$$y^{m+1} = y^0 + \int_0^t \{Q[x^m(s), y^m(s)] + E_2(s)\}\, ds.$$

If $t = T$, then the first two terms on the right-hand sides of these two equations depend only on x^0 and y^0. The last terms are constants which are independent of m. So we may conclude that the components x, y of $f_T(u^0)$ are of the form:

$$x = g_1(x^0, y^0) + k_1,$$

$$y = g_2(x^0, y^0) + k_2$$

where

$$k_i = \int_0^T E_i(s)\, ds \qquad (i = 1, 2).$$

Thus

$$f_T(u^0) - u^0 = (g_1(x^0, y^0) - x^0, g_2(x^0, y^0) - y^0) + (k_1, k_2).$$

Since the local degree at $\bar{0}$ and relative to σ of this mapping is d, the local degree at $(-k_1, -k_2)$ and relative to σ of the mapping

$$u'' = (g_1(x^0, y^0) - x^0, g_2(x^0, y^0) - y^0)$$

is also d by Theorem (6.7) of Chapter I. By Corollary (7.4) of Chapter I, we obtain the conclusion of the theorem if $E_1(t)$ and $E_2(t)$ are replaced by $E_1(t) + (a/\pi)\sin^2(2\pi t/T)$ and $E_2(t) + (b/\pi)\sin^2(2\pi t/T)$ (where a, b are arbitrarily small real numbers). ∎

Theorems (9.7) and (9.8) are similar to the Poincaré-Bendixson Theorem (9.3) in that they give rather general conditions under which periodic solutions exist and also in the fact that it is usually quite difficult to prove that these general conditions are satisfied.

Now we describe conditions under which Hypothesis (9.5) is satisfied. To prove that these sets of conditions imply Hypothesis (9.5) is a lengthy procedure and we shall omit the proofs. The first result is due to Gomory [**1**]. In stating the result we use the concept of critical points on the line at infinity. Introduced by Poincaré, the concept is described in detail by Lefschetz [**4**]. The conditions about critical points at infinity which we use have simple analytic criteria which will be described in lemmas following the main theorem.

(9.9) THEOREM. *Suppose that in the system*

$$\text{(9.4)} \qquad \begin{aligned} \dot{x} &= P(x, y) + E_1(t), \\ \dot{y} &= Q(x, y) + E_2(t) \end{aligned}$$

the functions $P(x, y)$ *and* $Q(x, y)$ *are polynomials of degree* n *in* x *and* y. *If the system*

$$\begin{aligned} \dot{x} &= P(x, y), \\ \dot{y} &= Q(x, y) \end{aligned}$$

has critical points at infinity and if there are no consecutive saddle points on the line at infinity then there is a 2-cell σ *and a function* $F(x, y)$ *such that*:

(1) $F(x, y)$ *has continuous first derivatives in* x *and* y;
(2) $F(x, y) > 0$ *for all* (x, y);
(3) $F(x, y) \equiv 1$ *in a circle which contains* σ;
(4) $\lim_{x^2+y^2\to 0} F(x, y) = 0$ *in such a way that the right-hand sides of*

$$\text{(9.4-F)} \qquad \begin{aligned} \dot{x} &= \{P(x, y) + E_1(t)\}F(x, y); \\ \dot{y} &= \{Q(x, y) + E_2(t)\}F(x, y) \end{aligned}$$

are bounded in the xy-plane and such that (9.4-F) *satisfies Hypothesis* (9.5).

Note that if we then apply Theorems (9.7) and (9.8) to (9.4-F), the periodic solutions of (9.4-F) that are obtained are contained in σ and hence by property (3) of $F(x, y)$, these are also periodic solutions of (9.4).

To facilitate the use of Theorem (9.9), we describe some analytic criteria about critical points on the line at infinity.

(9.10) LEMMA. *Let $P_n(x, y)$ and $Q_n(x, y)$ be the terms of degree n in $P(x, y)$ and $Q(x, y)$ respectively. Necessary and sufficient conditions that there exist critical points on the line at infinity are:*

(i) $P_n(x, 1) - xQ_n(x, 1)$ *does not vanish identically and has real factors, all of odd multiplicity;*

(ii) $Q_n(1, y) - yP_n(1, y)$ *does not vanish identically and has real factors, all of odd multiplicity;*

(iii) *the equations*

$$P_n(x, 1) - xQ_n(x, 1) = 0$$

and

$$Q_n(x, 1) = 0$$

have no common roots;

(iv) *the equations*

$$Q_n(1, y) - yP_n(1, y) = 0$$

and

$$P_n(1, y) = 0$$

have no common roots.

Lemma (9.10) is due to Gomory [**1**].

(9.11) LEMMA. *Let $y_c/x_c = \alpha$ be the ratio of the coordinates of a critical point at infinity, i.e., the x_c and y_c satisfy the equation*

$$xP_n(x, y) - yQ_n(x, y) = 0.$$

If as α increases from $\alpha - \varepsilon$ to $\alpha + \varepsilon$, the expression

$$\frac{Q_n}{P_n} - \frac{y}{x}$$

changes from negative to positive, the critical point is a saddle point. If the expression changes from positive to negative, it is a node.

Lemma (9.11) is due to Poincaré [**2**, p. 25].

(9.12) LEMMA. *Let $J(P_n, Q_n)$ denote the Jacobian of $P_n(x, y)$ and $Q_n(x, y)$. If $J(P_n, Q_n) < 0$ for all $(x, y) \neq (0, 0)$, there are no saddle sectors on the line at infinity.*

Lemma (9.12) is due to Cronin [**4**].

As a simple application of Theorem (9.9), we have:

(9.13) THEOREM. *If the conditions of Lemma (9.10) are satisfied and if $J(P_n, Q_n) < 0$ for all $(x, y) \neq (0, 0)$ then (9.4) has at least one solution of period*

T. If $E_1(t)$ and $E_2(t)$ are varied arbitrarily slightly (as described in the proof of Theorem (9.8)) then (9.4) has at least n distinct periodic solutions of period T and at most a finite number of solutions of period T.

PROOF. System (9.4) satisfies Hypothesis (9.5). Since $J(P_n, Q_n) < 0$ for all $(x, y) \neq (0, 0)$, then $d(M, \sigma, \bar{0})$ is $-n$. The conclusions of the theorem then follow from Theorems (9.7) and (9.8).

A simple example to which Theorem (9.13) may be applied is:

$$\dot{x} = y^n + p(x, y) + E_1(t),$$

$$\dot{y} = x^n + q(x, y) + E_2(t)$$

where n is any odd integer and $p(x, y)$ and $q(x, y)$ are polynomials of degree less than n.

Hypothesis (9.5) could be formulated for an n-dimensional system where $n > 2$. However, since only one simple result will be obtained here for $n > 2$, the hypothesis will not be stated formally. Consider the n-dimensional system where $n > 2$

$$x = X(x) + E(t) \tag{9.14}$$

where $E(t)$ is an n-vector each of whose components has a continuous first derivative and is of period T in t and each component of $X(x)$ has continuous first derivatives in $x_1, \cdots, x_n$ for all real $x_1, \cdots, x_n$ where $x_1, \cdots, x_n$ are the components of vector x. Assume that there is an n-cell σ in R^n such that if a solution $x(t)$ of (9.14) intersects σ', then the solution crosses σ' inwardly. Then the vector field

$$V_t(x^{(0)}) = x(t, x^{(0)}, t_0) - x^{(0)},$$

where $x(t, x^{(0)}, t_0)$ is the solution of (9.14) such that

$$x(t_0, x^{(0)}, t_0) = x^{(0)},$$

is such that the index of $V_t(x^{(0)})$ relative to σ is defined for all $t \in [0, T]$. By Theorem (13.1) of Chapter I, the index of $V_t(x^{(0)})$ relative to σ' is $+1$. Hence the mapping

$$f_T(x^0) = x(T, x^{(0)}, t_0)$$

has at least one fixed point in σ and thus (9.14) has at least one solution of period T. In particular, we have:

(9.15) THEOREM. *If there exist positive constants γ_1 and γ_2 such that if*

$$\sum_{1=1}^{n} x_i^2 \geqq \gamma_1,$$

then

$$\sum x_i X_i(x) \geqq \gamma_2$$

or

$$\sum x_i X_i(x) \leqq -\gamma_2;$$

then (9.14) *has at least one solution of period* T.

PROOF. If $\gamma = [\sum x_i^2 \, 2]^{1/2}$, then

$$\dot{r} = \frac{1}{r} \sum_{i=1}^{n} x_i X_i(x).$$

From the hypothesis, it follows that there exist positive constants r_3 and r_4 such that if $r > r_3$, then

$$\frac{1}{r} \sum x_i [X_i(x) + E_i(t)] > r_4 > 0$$

where $E_i(t)$ is the ith component of $E(t)$. The sphere in R^n with center at the origin and radius greater than r_3 is an n-cell σ which satisfies the condition described before the statement of Theorem (9.15). ∎

A simple example is:

$$\dot{x}_1 = x_1^m + P_1(x_1, \cdots, x_n) + E_1(t)$$

$$\cdots\cdots\cdots\cdots$$

$$\dot{x}_n = x_n^m + P_n(x_1, \cdots, x_n) + E_n(t)$$

where m is an odd integer and $P_i(x_1, \cdots, x_n)$ is a polynomial of degree less than m.

This n-dimensional result is similar to the following simple but often used application of the Poincaré-Bendixson Theorem (9.3): if the system

$$\dot{x} = P(x, y),$$

$$\dot{y} = Q(x, y)$$

has an unstable critical point at the origin and if there is a simple closed curve C containing the origin but no other critical point in its interior or on C itself and if every orbit which intersects C crosses C inwardly, then the system has a periodic solution. It is interesting to observe that here the nonautonomous case is easier to handle since we do not require that $n = 2$ and no hypothesis corresponding to the condition that the origin be an unstable critical point is required.

A far more significant result than Theorem (9.15) is the following result due to Levinson.

(9.16) THEOREM. *In the single second-order equation*

$$\ddot{x} + f(x, \dot{x})\dot{x} + g(x) = e(t)$$

which may be written as a first-order, 2-dimensional system

$$\begin{aligned} \dot{x} &= y, \\ \dot{y} &= -f(x, y)y - g(x) + e(t); \end{aligned} \tag{9.17}$$

let

$$G(x) = \int_0^x g(s)\, ds.$$

Assume the following conditions are satisfied:

1. *for all x, y, the functions $f(x, y)$ and $g(x)$ are continuous and each satisfies a Lipschitz condition;*
2. *for x large $g(x)$ has the sign of x and $|g(x)|$ increases monotonically to infinity with $|x|$ and there is a positive α such that*

$$\frac{\dfrac{g(x)}{G(x)}}{\dfrac{1}{|x|}} < \alpha$$

for $|x|$ sufficiently large;

3. *there exists an $a > 0$ such that $f(x, y) \geqq M > 0$ for $|x| \geqq a$ and $f(x, y) \geqq -m$, $m > 0$ for $|x| \leqq a$;*
4. *function $e(t)$ is continuous and there is a positive constant E such that $|e(t)| \leqq E$ for all t.*

Then there is a 2-cell σ such that each solution of (9.17) *that intersects σ' crosses σ' inwardly.*

The proof of Theorem (9.16) is lengthy. See Levinson [**1**] and Lefschetz [**4**, Chapter XI]. Because of its importance in applications, this theorem has been extended in various directions by many writers.

From the discussion preceding Theorem (9.15) we have:

(9.18) Corollary. *If $e(t)$ has period T, then system* (9.17) *has a solution of period T.*

As shown by Levinson [**1**] and Lefschetz [**4**], Corollary (9.18) can also be proved simply by using the Brouwer Fixed Point Theorem (Theorem (1.2) of Chapter I).

CHAPTER III

Topological Techniques in Function Space

1. **Introduction.** In this chapter, the topological techniques of Chapter I are extended from finite-dimensional spaces to function space. "Function space" is a vague term which usually refers to infinite-dimensional Banach spaces and Hilbert spaces (the latter concepts will be precisely defined in this chapter) because the various linear spaces of functions which occur in analysis are among the most important concrete examples of these abstract spaces. Extending the techniques from the finite-dimensional to the infinite dimensional spaces presents a serious problem. The basic reason for this is that the unit sphere (more generally any closed bounded set) in the finite-dimensional case is compact while the unit sphere in the infinite-dimensional case is not compact. Because of this it is impossible to obtain a fixed point theorem and a local degree for arbitrary continuous mappings. Instead only a special class of mappings which have a compactness property can be considered. For these, we prove the Schauder Fixed Point Theorem (Schauder [**1**]) and a striking extension by Schaefer [**1**] of that theorem; for related mappings, the Leray-Schauder degree is defined and its basic properties derived (Leray and Schauder [**1**]).

Next we take a more analytic viewpoint and prove the Banach Fixed Point Theorem for contraction mappings by using successive approximations. Finally, we consider a kind of combination of the Leray-Schauder theory and the Banach Theorem and obtain detailed local information about mappings of the form $I + C + T$ where $I + C$ is a linear mapping of the type studied by Leray and Schauder and T is a "higher-order" mapping.

Although applications of this theory will not be described until Chapter IV, we indicate here the nature of these applications in order to clarify the status of the various theorems in this chapter. Probably the most frequently used technique in applications is the Schauder Fixed Point Theorem. But the strongest and most interesting applications use the Leray-Schauder degree theory. Actually in all applications of the degree theory, only a particular corollary of the Leray-Schauder theory is used. Schaefer [**1**] has given a simple proof of this corollary which requires no use of the Leray-Schauder degree itself. Thus the Leray-Schauder degree itself is never used in applications. The reason for this may be described in topological terms by saying that all the mappings studied in the applications have Leray-Schauder degree $+1$. This suggests the possibility of applying the Leray-Schauder degree to functional equations in which the degree of corresponding mapping is different from $+1$. However, except in local studies (§ 8 of Chapter III and § 10 of Chapter IV) no such applications have been made thus far.

We will deal with mappings from a normed linear space into itself since most applications to analysis (up to the present) are made in these spaces, but much of the theory of this chapter has been extended to a wider class of linear spaces. The Schauder Fixed Point Theorem was generalized to mappings in a locally convex linear topological space by Tychonoff [**1**] and certain extensions to arbitrary linear topological spaces have been made by Klee [**2**; **3**]. Extensions in a different direction have been made by Browder [**1**] who proved fixed point theorems for mappings in a Banach space such that a power of the mapping has the property that the mapping itself is required to have in the Schauder Fixed Point Theorem and mappings from points in a Banach space into closed convex subsets of the Banach space. (Browder remarks that his results are valid also in locally convex linear topological spaces.) The Leray-Schauder degree theory has been extended to mappings in locally convex linear topological spaces by Nagumo [**2**] and to arbitrary linear topological spaces by Klee [**2**; **3**].

2. **Some linear space theory.** The framework within which the Schauder Fixed Point Theorem will be proved and the Leray-Schauder degree defined is the normed linear space. For the Banach Fixed Point Theorem and for the local study of the transformation $I + C + T$, we will need a complete normed linear space, i.e., a Banach space.

DEFINITION. Let R denote the field of real numbers or the field of complex numbers. A *linear space L over R* is a set of elements $x, y, \cdots$ satisfying the following conditions:

(1) L is a commutative group (group operation denoted by $+$ and identity denoted by $\bar{0}$);

(2) corresponding to each pair $a \in R$, $x \in L$, there is an element of L, called the product of a and x and denoted by ax, such that

(i) $a(x + y) = ax + ay$;

(ii) $(a + b)x = ax + bx$;

(iii) $a(bx) = (ab)x$;

(iv) $0x = \bar{0}$ and $1x = x$.

(We will deal in the applications only with linear spaces over the real numbers.)

DEFINITION. A set of elements $x_1, \cdots, x_n$ in a linear space L is *linearly independent* if $a_1x_1 + \cdots + a_nx_n = \bar{0}$, where $a_i \in R$ for $i = 1, \cdots, n$, implies that $a_1 = a_2 = \cdots = a_n = 0$.

DEFINITION. A subset K of linear space L is a *linear manifold of L* if $x, y \in K$ implies that $ax + by \in K$ for all $a, b \in R$. (Note that L is a linear manifold of L.)

Definition. If $[x_\nu]$ is a subset of linear space L, the *linear manifold of L spanned by* $[x_\nu]$ is $\bigcap_\mu L_\mu$ where L_μ is a linear manifold of L such that $[x_\nu] \subset L_\mu$ and the intersection is taken over all such L_μ. (It is easy to show that $\bigcap_\mu L_\mu$ is a linear manifold of L.)

Definition. If for each positive integer m, there is in linear manifold K a set of m linearly independent vectors then K is *infinite-dimensional*.

Definition. If n is the maximum number of elements in a linearly independent set in K, then K is *finite-dimensional* and has *dimension n* or is *n-dimensional*.

Definition. If $x_1, \cdots, x_n$ is a set of n linearly independent elements in linear manifold K such that each element y of K can be expressed as a linear combination of the x_i's, i.e.,

$$y = \sum_{i=1}^{n} b_i x_i,$$

then $x_i, \cdots, x_n$ is a *basis* of K.

It is easy to show that an n-dimensional linear manifold has a basis consisting of n elements.

Definition. A *norm* on a linear space L is a mapping from L into the non-negative real numbers, with the norm value corresponding to $x \in L$ denoted by $\|x\|$, and such that:

(1) if $x \neq \bar{0}$, then $\|x\| > 0$;

(2) $\|ax\| = |\alpha|\ \|x\|$;

(3) $\|x + y\| \leqq \|x\| + \|y\|$.

Definition. A *normed linear space* is a linear space with a norm.

Definition. A set E in a normed linear space $\mathcal{N}$ is *bounded* if there is a positive number A such that if $x \in E$, then $\|x\| \leqq A$.

Theorem. *If $\mathcal{N}$ is a normed linear space, then*

$$\rho(x, y) = \|x - y\|$$

is a metric under which $\mathcal{N}$ is a metric space.

Proof. Using the definition of norm, it is easy to prove that $\rho(x, y)$ satisfies the conditions for a metric:

(i) $\rho(x, y) > 0$ if and only if $x \neq y$;

(ii) $\rho(x, y) = \rho(y, x)$;

(iii) $\rho(x, y) \leqq \rho(x, z) + \rho(z, y)$.

We say that the metric $\rho(x, y)$ is *induced* by the norm.

Definition. A *subspace of $\mathcal{N}$* is a linear manifold of $\mathcal{N}$ which is a closed subset of $\mathcal{N}$.

DEFINITION. If $[x_\nu]$ is a subset of a normed linear space $\mathcal{N}$, the *subspace of $\mathcal{N}$ spanned by* $[x_\nu]$ is $\bigcap_\mu L_\mu$ where L_μ is a subspace of $\mathcal{N}$ such that $[x_\nu] \subset L_\mu$ and the intersection is taken over all such L_μ. (It is easy to show that $\bigcap_\mu L_\mu$ is a subspace of $\mathcal{N}$.)

DEFINITION. A *sphere of radius r and center x_0* in a normed linear space $\mathcal{N}$ is the set:

$$[x \in \mathcal{N} \;/\; \|x - x_0\| \leqq r].$$

DEFINITION. A metric space M is *complete* if for every sequence $\{x_n\}$ in M which satisfies a Cauchy condition (in terms of the metric) there is a limit in M, i.e., an element x of M such that

$$\lim_{n \to \infty} \rho(x_n, x) = 0.$$

DEFINITION. A *Banach space* is a normed linear space which is complete in the metric induced by the norm. (The topology of the Banach space or normed linear space given by this metric is sometimes called the norm topology. There are other important topologies on the Banach space. Since none of these topologies will be used here, we omit description of them.)

DEFINITION. An *inner product* ϕ on a linear space L is a mapping from the set of ordered pairs (x, y) where $x, y \in L$, into the complex numbers such that:

(i) if $x \neq \bar{0}$, then $\phi(x, x) > 0$;

(ii) $\phi(x, y) = [\phi(y, x)]^*$, where $*$ denotes the conjugate of a complex number;

(iii) $\phi(ax + by, z) = a\phi(x, z) + b\phi(y, z)$ and $\phi(x, by + cz) = b^*\phi(x, y) + c^*\phi(x, z)$.

DEFINITION. An *inner product* space is a linear space with an inner product. It can be proved that $(x, x)^{1/2}$ is a norm which in turn induces a metric. We say that the inner product induces a norm and a metric.

DEFINITION. A *Hilbert space* is an inner product space which is complete in the metric induced by the inner product.

DEFINITION. If x and y are elements of S, an inner product space, then x is *orthogonal* to y if $(x, y) = 0$.

DEFINITION. A set $[x_j]$ in S, an inner product space, is an *orthonormal* set if $(x_i, x_j) = \delta_{ij}$ for all i, j.

DEFINITION. A *basis* for an infinite-dimensional Hilbert space $\mathscr{H}$ is a set $[x_j]$ $(j = 1, 2, \cdots)$ such that for each $x \in \mathscr{H}$ there is a sequence of numbers $\{a_j\}$ such that

(L)
$$\lim_{n = \infty} \left\| x - \sum_{j=1}^{n} a_j x_j \right\| = 0.$$

We express the fact that condition (L) holds by writing: $x = \sum_{j=1}^{\infty} a_j x_j$.

(2.1) THEOREM. *If $\mathscr{H}$ is a separable infinite-dimensional Hilbert space, then $\mathscr{H}$ has an orthonormal basis and if $x = \sum_{j=1}^{\infty} a_j x_j$, then*

$$\sum_{j=1}^{\infty} |a_j|^2 \leqq \|x\|^2.$$

(*This inequality is called Bessel's inequality.*)

Since we need this theorem just for one example, we omit the proof which may be found in Stone [**1**]. If the nonseparable case is included, a basis may also be obtained but it is not, in general, a denumerable set and the limiting process must be defined differently. See Halmos [**1**, Chapter 1].

From the definitions, it is clear that a Hilbert space is a Banach space. But there are many Banach spaces, important in applications, which are not Hilbert spaces. For both the Hilbert space and the Banach space, there is an extensive theory (see Halmos [**1**], Banach [**1**], Riesz and Nagy [**1**], and Day [**1**]) most of which will not be mentioned here. But to clarify the difference between a Banach space and a Hilbert space, we state the following theorem:

(2.2) THEOREM. *A nasc that Banach space $\mathscr{B}$ be a Hilbert space is that for each pair $f, g \in \mathscr{B}$, it is true that*

$$\|f + g\|^2 + \|f - g\|^2 = 2\|f\|^2 + 2\|g\|^2.$$

(*This equality is sometimes called the parallelogram equality.*)

For proof, see Jordan and von Neumann [**1**].

EXAMPLES. The most familiar example of a Hilbert space is the n-dimensional Euclidean space R^n of n-tuples $x = (x_1, \cdots, x_n)$ with the inner product

$$(x, y) = \sum_{i=1}^{n} x_i y_i^*.$$

The simplest infinite-dimensional Hilbert space is l^2, the linear space of square summable sequences of real or complex numbers, i.e., sequences $x = \{x_i\}$ such that

$$\sum_{i=1}^{\infty} |x_i|^2 < \infty.$$

The inner product is

$$(x, y) = \sum x_i y_i^*.$$

Another frequently studied Hilbert space is the space of square-integrable functions on $[0, 1]$, i.e., the linear space of functions f such that the Lebesgue integral $\int_{[0,1]} |f|^2$ is finite. The inner product is $(f, g) = \int_{[0,1]} f(g)^*$.

Finally a simple example of a Banach space is the linear space of real-valued continuous functions f on $[0, 1]$ with the norm:

$$\|f\| = \max_{x \in [0,1]} |f(x)|.$$

By Theorem (2.2), this Banach space is not a Hilbert space because if we consider:

$$f(x) = (1 - 2x)^2, \qquad 0 \leqq x \leqq \tfrac{1}{2},$$
$$f(x) = 0, \qquad \tfrac{1}{2} < x \leqq 1$$

and

$$g(x) = 0, \qquad 0 \leqq x \leqq \tfrac{1}{2},$$
$$g(x) = 2(1 - x)^2 - 1, \qquad \tfrac{1}{2} < x \leqq 1,$$

the parallelogram equality does not hold for this pair of elements of the Banach space.

DEFINITION A *transformation* or *operator* from a normed linear space $\mathcal{N}$ into a normed linear space $\mathcal{N}_1$ is a function the domain of which is a subset of $\mathcal{N}$ and the range of which is a subset of $\mathcal{N}_1$.

We shall deal only with transformations from a normed linear space $\mathcal{N}$ into itself which are continuous in the norm topology of $\mathcal{N}$.

DEFINITION. Transformation T from $\mathcal{N}$ into $\mathcal{N}$ is *linear* if T is continuous and if for all numbers a, b and all $x, y \in \mathcal{N}$:

$$T(ax + by) = aT(x) + bT(y).$$

If T is a linear transformation, then

$$\operatorname*{lub}_{\|x\|=1} \|T(x)\| = \operatorname*{lub}_{x \in \mathscr{B}} \frac{\|T(x)\|}{\|x\|}$$

is finite. We call this number the norm of T and denote it by $\|T\|$. It can be proved that this norm has the usual norm properties and that under it the linear space of linear transformations T is a Banach space if $\mathcal{N}$ and $\mathcal{N}_1$ are Banach spaces.

3. **Examples which show that a fixed point theorem and a definition of local degree cannot be obtained for arbitrary continuous transformations from a Banach space into a Banach space.** Now we are ready to prove a fixed point theorem and define a local degree for mappings in a normed linear space. It is natural at the outset to assume that our mappings are simply continuous as in the finite-dimensional case (Chapter I). However, we describe now two examples which show that it is impossible to obtain a fixed point theorem or a local degree for arbitrary continuous mappings. We will then prove a fixed point theorem and define a local degree for a special class of mappings which satisfy a strong compactness condition. (Klee [**1**] has shown that this compactness condition is also necessary. Klee proved that if E is a locally convex metrizable linear topological space and K is a non-compact convex subset of E, then K lacks the fixed point property, i.e., there is a continuous mapping of K into itself which does not have a fixed point.)

KAKUTANI'S EXAMPLE. This is an example of a homeomorphism of the unit sphere of a Hilbert space onto itself which has no fixed point. See Kakutani [**1**].

Let $\mathscr{H}$ be a separable Hilbert space and let B be the unit sphere with center $\overline{0}$, i.e.,

$$B = [x \in \mathscr{H} \mid \|x\| \leqq 1].$$

Let $\{y_n\}$ $(n = 0, \pm 1, \pm 2, \cdots)$ be an orthonormal basis for $\mathscr{H}$ and define the transformation U this way: let

$$U : y_n \rightarrow y_{n+1} \qquad (n = 0, \pm 1, \pm 2, \cdots)$$

and extend the definition of U linearly, i.e., for arbitrary $x \in \mathscr{H}$, where $x = \sum a_i y_i$, define

$$U(x) = \sum a_i U(y_i) = \sum a_i y_{i+1},$$

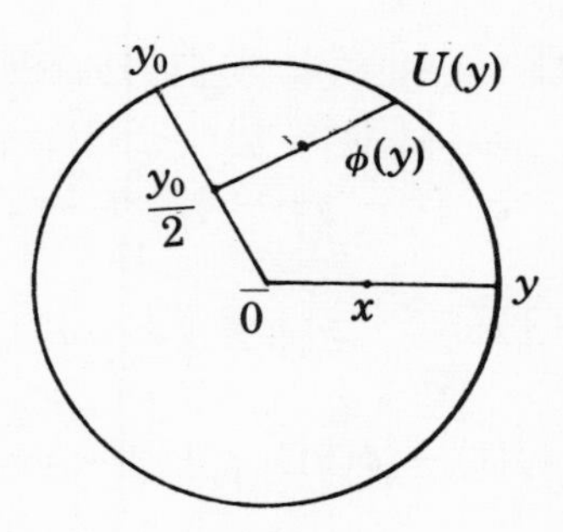

FIGURE 1

so that the domain of U is $\mathscr{H}$. It follows that U is a homeomorphism of $\mathscr{H}$ onto itself. Also U is a homeomorphism of the set

$$S = [x \in \mathscr{H} \mid \|x\| = 1]$$

onto itself. Now define the mapping ϕ on $\mathscr{H}$ as:

$$\phi : x \rightarrow \tfrac{1}{2}(1 - \|x\|)y_0 + U(x).$$

We show that ϕ is a homeomorphism. First ϕ is continuous because U is continuous. To show that ϕ is 1-1, we write, for $x \neq \overline{0}$,

$$\phi(x) = (1 - \|x\|)\frac{y_0}{2} + \|x\|[U(y)]$$

where $y = x/\|x\|$. This shows that $\phi(x)$ divides the line segment joining $\frac{1}{2}y_0$ and $U(y)$ in the same proportion as the point x divides the segment joining $\overline{0}$ and y. (Figure 1.)

From this we see that if $x^{(1)} \neq x^{(2)}$, then $\phi(x^{(1)}) \neq \phi(x^{(2)})$. For if $x^{(1)}$ and $x^{(2)}$ are linearly dependent, then $\phi(x^{(1)})$ and $\phi(x^{(2)})$ are different points on the same line segments; if $x^{(1)}$ and $x^{(2)}$ are linearly independent then

$$U\left(\frac{x^{(1)}}{\|x^{(1)}\|}\right) \quad \text{and} \quad U\left(\frac{x^{(2)}}{\|x^{(2)}\|}\right)$$

are linearly independent, hence distinct. Thus $\phi(x^{(1)})$ and $\phi(x^{(2)})$ are on different line segments and are distinct. To show that ϕ^{-1} is continuous, note that for all x,

$$\|U(x)\| = \|x\|.$$

Also from condition 3 in the definition of norm, it follows that for all x and y,

$$|\,\|x\| - \|y\|\,| \leqq \|x - y\|.$$

Therefore

$$\begin{aligned} \|\phi(x) - \phi(y)\| &= \|\tfrac{1}{2}(-\|x\| + \|y\|)y_0 + U(x - y)\| \\ &\geqq |\,\|U(x - y)\| - \|\tfrac{1}{2}(\|x\| - \|y\|)y_0\|\,| \\ &\geqq \|x - y\| - \tfrac{1}{2}|\,\|x\| - \|y\|\,|. \end{aligned}$$

Since

$$|\,\|x\| - \|y\|\,| \leqq \|x - y\|,$$

then

$$\|\phi(x) - \phi(y)\| \geqq \tfrac{1}{2}\|x - y\|.$$

The continuity of ϕ^{-1} follows from this inequality.

Now we show that ϕ/B does not have a fixed point. Suppose there exists $x_0 \in B$ such that:

$$x_0 = \phi(x_0).$$

Since

$$x_0 = \phi(x_0) = \tfrac{1}{2}(1 - \|x\|)y_0 + U(x_0),$$

we have:

$$x_0 - U(x_0) = \tfrac{1}{2}(1 - \|x_0\|)y_0. \tag{3.1}$$

If $x_0 = \overline{0}$, then

$$\tfrac{1}{2}(1 - \|x_0\|)y_0 = \overline{0}.$$

Since $y_0 \neq \overline{0}$, then $1 - \|x_0\| = 0$ or $\|x_0\| = 1$. This contradicts our assumption that $x_0 = \overline{0}$. If $\|x_0\| = 1$, then $x_0 = U(x_0)$. But U does not have a fixed point on S because if U has a fixed point on S, then $\{y_n\}$ is not an orthonormal basis. Thus

$$0 < \|x_0\| < 1.$$

Since $\{y_n\}$ is an orthonormal basis, then by Theorem (2.1),

$$x_0 = \sum_{n=-\infty}^{\infty} a_n y_n$$

where $\sum_{n=-\infty}^{\infty} |a_n|^2 \leqq \|x_0\|^2 < 1$. But

$$x_0 - U(x_0) = \sum_{n=-\infty}^{\infty} (a_n - a_{n-1}) y_n. \tag{3.2}$$

Since $\{y_n\}$ is an orthonormal basis, then from (3.1) and (3.2), we have

$$a_0 - a_{-1} = \tfrac{1}{2}(1 - \|x\|) > 0$$

and

$$a_n = a_{n-1} \text{ for } n \neq 0.$$

Thus

$$a_{-3} = a_{-2} = a_{-1} < a_0 = a_1 = a_2 = \cdots.$$

But this contradicts the fact that (by Theorem (2.1))

$$\sum_{n=-\infty}^{\infty} |a_n|^2 < \infty.$$

Hence ϕ does not have a fixed point.

This example shows that it is impossible to prove a fixed point theorem in a Banach space which is valid for arbitrary continuous mappings. We may also use this example to show that it is impossible to define a local degree for arbitrary continuous mappings in a Banach space. First there is a continuous mapping ψ of B onto S such that if $x \in S$, then $\psi(x) = x$. The mapping ψ is defined this way: let $\psi(x)$ be the point in which the extension of the vector $(\phi(x), x)$ meets S. Since $\phi(x)$ has no fixed points, then ψ is defined for every $x \in S$. Since ϕ is continuous, the mapping ψ is continuous. Also if $x \in S$, then $\psi(x) = x$.

From this, it follows that the identity mapping $\phi_1(x) = x$ is homotopic in S to the constant mapping

$$\phi_0 : x \rightarrow \psi(\bar{0}),$$

i.e., there is a continuous mapping Ψ from $B \times [0, 1]$ into S such that for all $x \in S$,

$$\Psi(x, 0) = \psi(\bar{0})$$

and

$$\Psi(x, 1) = x.$$

We simply define $\Psi(x, t) = \psi(tx)$.

This shows that it is impossible to construct a local degree theory for arbitrary continuous mappings in a Banach space because in a local degree theory we expect the following conditions to be satisfied:

(i) the local degree of the identity mapping relative to $\bar{\omega}$, the closure of a bounded open set, at any point $x \in \omega$, is $+1$;

(ii) the local degree of a constant mapping (with range the point v) relative to $\bar{\omega}$ at any point x such that $x \neq v$ is 0;

(iii) if mappings M_1 and M_2 are defined on $\bar{\omega}$ and if there is a homotopy $H(x, t)$ with domain $\bar{\omega} \times [0, 1]$ such that

$$H(x, 0) = M_1(x),$$
$$H(x, 1) = M_2(x)$$

and such that for a fixed y we have $y \neq H(x, t)$ for all $(x, t) \in \omega' \times [0, 1]$, then the local degrees of M_1 and M_2 relative to $\bar{\omega}$ and at y are equal.

Leray's example. This is an example which shows the impossibility of defining a local degree for an arbitrary continuous mapping of a Banach space into itself. See Leray [**2**].

Let $\mathscr{B}$ be the Banach space of continuous functions $x(s)$ on $[0, 1]$ with the norm:

$$\|x(s)\| = \max_{s \in [0,1]} |x(s)|.$$

Let

$$x_0(s) \equiv \tfrac{1}{2}$$

and let

$$D = [x(s) \in \mathscr{B} \,/\, \|x - x_0\| < \tfrac{1}{2}].$$

Thus D is a bounded open set which consists of the functions $x(s)$ such that

$$0 < x(s) < 1$$

for all $s \in [0, 1]$. We define a mapping Φ taking $\bar{D}$ into $\mathscr{B}$ in the following way: let $\phi(x)$ be a real-valued continuous function with domain $[0, 1]$ such that for all x,

$$0 \leqq \phi(x) \leqq 1$$

and such that

$$\phi(0) = 0$$

and

$$\phi(1) = 1.$$

Then

$$\Phi : x(s) \to \phi[x(s)]$$

is a continuous mapping from $\bar{D}$ into $\bar{D}$. The transformation

$$H[x, t] = t\Phi[x] + (1 - t)[x] \qquad (t \in [0, 1])$$

is a homotopy such that $H[x, 0]$ is I, the identity mapping and $H[x, 1]$ is Φ.

Also if $y(s) \in D' = \bar{D} - D$, then for each $t \in [0, 1]$,

$$H[y, t] \in D'$$

because if $y(s) \in D'$, then $\|x - y\| = \frac{1}{2}$. Hence for all $s \in [0, 1]$,

$$0 \leqq y(s) \leqq 1,$$

and there is an $s_0 \in [0, 1]$ such that

$$y(s_0) = 0$$

or

$$y(s_0) = 1.$$

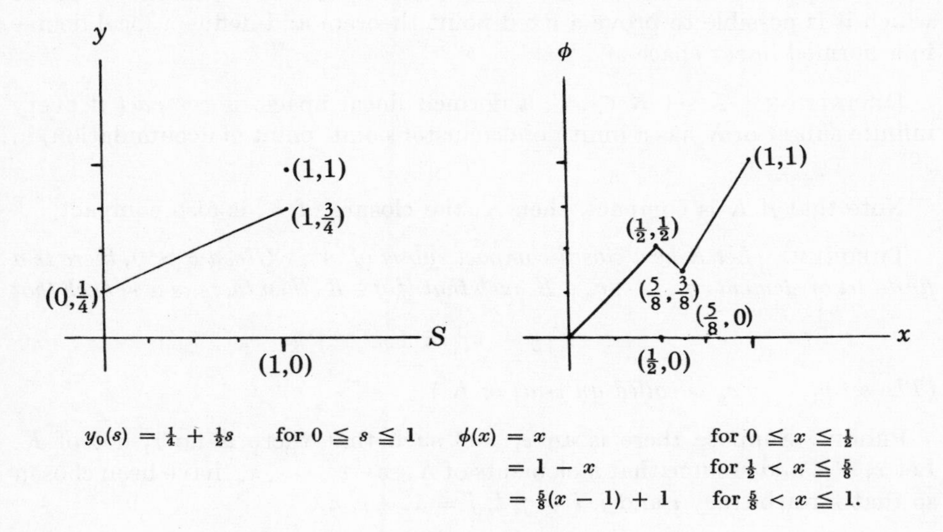

But the same conditions hold for $H[y, t]$ because from the definition of Φ, the expression

$$t\phi[y(s_0)] + (1 - t)y(s_0)$$

is 0 for all $t \in [0, 1]$ if $y(s_0) = 0$ and is 1 for all $t \in [0, 1]$ if $y(s_0) = 1$. Also for arbitrary $s \in [0, 1]$ and for all $t \in [0, 1]$

$$0 \leqq \{t\phi[y(s)] + (1 - t)y(s)\} \leqq t + (1 - t) = 1.$$

Thus if $y(s) \in D'$, then $H[y, t] \in D'$ for all $t \in [0, 1]$. Hence if $y_0 \in D$, mapping Φ is homotopic to I in $\mathscr{B} - [y_0]$, i.e., the space $\mathscr{B}$ with the element y_0 deleted. Hence if a local degree can be defined, we should obtain: the local degree of Φ relative to $\bar{D}$ at y_0 is $+1$ and by the basic property of local degree, the equation

$$\Phi[x(s)] = y_0(s) \tag{3.3}$$

should have a solution $x(s)$ in D. But this is not generally true as the following example shows.

Let $y_0(s)$ and $\phi(x)$ be defined as on page 129.

Since

$$\phi[x(0)] - y(0) = \tfrac{1}{4}$$

then $x(0) = \frac{1}{4}$. But $y_0(s)$ increases monotonically from $\frac{1}{4}$ to $\frac{3}{4}$, an increase of $\frac{1}{2}$, while $\phi[x(s)]$ can increase monotonically (from $s = 0$) at most $\frac{1}{4}$ and then decrease. Thus equation (3.3) does not have a solution $x(s)$ in this case.

4. **Compact transformations.** Now we impose the conditions under which it is possible to prove a fixed point theorem and define a local degree in a normed linear space $\mathscr{N}$.

DEFINITION. A set $K \subset \mathscr{N}$, a normed linear space, is *compact* if every infinite subset of K has a limit point (cluster point, point of accumulation) in $\mathscr{N}$.

Note that if K is compact, then $\bar{K}$, the closure of K, is also compact.

THEOREM. *Let K be a closed compact subset of $\mathscr{N}$. Given $\varepsilon > 0$, there is a finite set of elements $v_1, \cdots, v_p \in K$ such that if $y \in K$, then there is a v_i such that*

$$\|y - v_i\| < \varepsilon.$$

(The set $v_1, \cdots, v_p$ is called an ε-net of K.)

PROOF. Suppose there is an $\varepsilon_1 > 0$ such that there is no ε_1-net of K. Let $x_1 \in K$ and assume that n elements of K, say $x_1, \cdots, x_n$, have been chosen so that for arbitrary i and j, $i \neq j$, $i, j = 1, \cdots, n$,

$$\|x_i - x_j\| \geqq \varepsilon_1.$$

Since there is no ε_1-net of K, there is an element x_{n+1} of K such that

$$\|x_{n+1} - x_j\| \geqq \varepsilon_1$$

for $j = 1, \cdots, n$. Thus we define inductively an infinite subset $[x_n]$ of K.

Now let y be an arbitrary element of $\mathscr{N} - [x_n]$ and let

$$N_{\varepsilon_1}(y) = \left[z \,/\, \|z - y\| < \frac{\varepsilon_1}{2}\right].$$

There can be at most one point of $[x_n]$ in $N\varepsilon_1(y)$. Hence y cannot be a limit point of $[x_n]$. Thus $[x_n]$ is an infinite subset of K and $[x_n]$ has no limit point. Therefore K is not compact. So our original supposition that there is no ε_1-net of K must be false. ▮

DEFINITION. Let E be a subset of $\mathcal{N}$. Transformation

$$T : E \to \mathcal{N}$$

is *compact* or *completely continuous* if:

(i) T is continuous;

(ii) if M is a bounded subset of E, then $T(M)$ is compact.

Now let K be a compact set and $\bar{K}$ its closure. Let $v_1, \cdots, v_p$ be an ε-net of $\bar{K}$. For $x \in \bar{K}$, define:

$$F_\varepsilon(x) = \frac{\sum_{i=1}^{n} m_i(x) v_i}{\sum_{i=1}^{n} m_i(x)}$$

where:

$$m_i(x) = \varepsilon - \|x - v_i\| \qquad \text{if } \|x - v_i\| \leqq \varepsilon,$$
$$m_i(x) = 0 \qquad \text{if } \|x - v_i\| > \varepsilon.$$

THEOREM. *Let T be a compact transformation with domain M, a bounded subset of $\mathcal{N}$, and let $T(M) \subset K$. Let F_ε be defined on $\bar{K}$ as described above. If $x \in M$, then*

$$\|T(x) - F_\varepsilon T(x)\| < \varepsilon.$$

PROOF.

$$\|T(x) - F_\varepsilon T(x)\| = \frac{\left\| \sum_{i=1}^{p} m_i[T(x)]T(x) - \sum_{i=1}^{p} m_i[T(x)]v_i \right\|}{\sum_{i=1}^{p} m_i[T(x)]}$$

$$\leqq \frac{\sum_{i=1}^{p} m_i[T(x)] \| T(x) - v_i \|}{\sum_{i=1}^{p} m_i[T(x)]} < \varepsilon. \blacksquare$$

DEFINITION. If $E \subset \mathcal{N}$ and if $x, y \in E$ implies $ax + by \in E$ for $0 \leqq a \leqq 1$, $0 \leqq b \leqq 1$, and $a + b = 1$, then E is *convex*.

(4.1) SCHAUDER FIXED POINT THEOREM. *Let K be a convex closed set and T a compact transformation such that $T(K) \subset K$. Then T has a fixed point, i.e., there is an $x \in K$ such that $T(x) = x$.*

PROOF. First since K is closed, the set $\overline{T(K)}$, the closure of $T(K)$, is contained in K. Let $\{\varepsilon_n\}$ be a monotonic decreasing sequence such that

$$\lim_{n \to \infty} \varepsilon_n = 0.$$

Let $T_n = F_{\varepsilon_n} T$ be defined on K as described in the previous theorem. Since K is convex, then

$$T_n(K) \subset K$$

because

$$F_{\varepsilon_n}(x) = \frac{\sum_{i=1}^{p_n} m_i(x) v_i}{\sum_{i=1}^{p_n} m_i(x)}$$

where $v_1, \cdots, v_{p_n}$ is an ε_n-net of $\overline{T(K)}$ and $\overline{T(K)} \subset K$.

Let $\mathscr{N}_n$ be the finite-dimensional subspace of $\mathscr{N}$ which is spanned by $v_1, \cdots, v_{p_n}$. Let

$$K_n = K \cap \mathscr{N}_n.$$

This is a closed convex subset of $\mathscr{N}_n$. The transformation T_n is defined on K_n and

$$T_n(K_n) \subset K_n.$$

Hence by Corollary (11.2) of the fixed point theorem in Chapter I, there is a point $x_n \in K_n$ such that

$$T_n(x_n) = x_n.$$

The set $[T(x_n)]$ is contained in the closed compact set $\overline{T(K)}$ and hence $[T(x_n)] \subset K$. Thus $[T(x_n)]$ has a limit point x_0 and $x_0 \in K$. Either sequence $\{T(x_n)\}$ converges to x_0 or there is a subsequence of $\{T(x_n)\}$ which converges to x_0. For simplicity of notation, assume $\{T(x_n)\}$ converges to x_0.

Then

$$|T(x_n) - x_0| < \varepsilon \qquad \text{for } n > n(\varepsilon). \tag{4.2}$$

Also from the definition of T_n, we have:

$$|T_n(x_n) - T(x_n)| < \varepsilon_n. \tag{4.3}$$

Adding (4.2) and (4.3), we have:

$$|T_n(x_n) - x_0| < \varepsilon + \varepsilon_n.$$

Since

$$T_n(x_n) = x_n,$$

then

$$|x_n - x_0| < \varepsilon + \varepsilon_n.$$

Since T is continuous then if $\varepsilon + \varepsilon_n < \delta(\varepsilon^{(1)})$,

$$|T(x_n) - T(x_0)| < \varepsilon^{(1)}.$$

Thus the sequence $\{T(x_n)\}$ converges to $T(x_0)$. Since the limit of the sequence $\{T(x_n)\}$ is unique, then

$$x_0 = T(x_0). \blacksquare$$

COROLLARY. *Let K be a homeomorph of a convex closed set and T a compact transformation such that $T(K) \subset K$. Then T has a fixed point.*

PROOF. The proof follows from Theorem (4.1) by using the fact that if T is a compact mapping and H is a homeomorphism, then $H^{-1}TH$ is a compact mapping.

We could have avoided the hypothesis of convexity by obtaining the result for a homeomorph of the unit sphere (the unit sphere is, of course, a convex set) as in the finite-dimensional case (Chapter I). The disadvantage of this method in the infinite-dimensional case is that the characterization of convex sets is far more complex than in the finite-dimensional case (see Klee [**4**; **5**]) and an analog of Corollary (11.2), in Chapter I would not be easy to prove as was Corollary (11.2).

In 1922, Birkhoff and Kellogg [**1**] set up a technique for proving the existence of fixed points of mappings in function spaces and applied it to mappings in two particular spaces. Schauder's Theorem may be regarded as a generalization of the Birkhoff-Kellogg result. The important contribution of the Schauder paper is its explicit use of the concept of compactness.

Now we obtain an important extension of the fixed point theorem which was originally proved by using the Leray-Schauder degree of mappings in a normed linear space. The simple proof obtained by Schaefer [**1**] is given here. Schaefer proves the theorem in a locally convex linear topological space. The proof given here, which is almost identical, is phrased in the more restrictive framework of a normed linear space.

(4.4) SCHAEFER'S THEOREM. *Let T be a compact transformation of a normed linear space $\mathcal{N}$ into itself. Let $\lambda_1 \in [0, 1]$. Then either there is an $x \in \mathcal{N}$ such that*

$$x = \lambda_1 T(x)$$

or the set

$$\{x \in \mathcal{N} \mid x = \lambda T(x),\ 0 < \lambda < 1\}$$

is not bounded.

PROOF. Let $U = \{x \mid \|x\| \leqq 1\}$;

$$nU = \{y \mid y = nx \text{ where } x \in U\}$$

where n is a positive integer.

Suppose there is a $\lambda_0 \in (0, 1)$ such that the equation

$$x = \lambda_0 T(x)$$

does not have a solution. We prove: given positive integer n there is an x_n such that

$$x_n = \mu_n \lambda_0 T(x_n),$$

where $\mu_n \in (0, 1)$ and $\|x_n\| = n$. Define the mapping $R_n: \mathcal{N} \to \mathcal{N}$ as follows:

$$R_n(x) = \lambda_0 T(x) \qquad \text{for } x \text{ such that } \lambda_0 T(x) \in nU;$$

$$R_n(x) = \frac{n}{\|\lambda_0 T(x)\|} \lambda_0 T(x) \qquad \text{for } x \text{ such that } \lambda_0 T(x) \in \mathcal{N} - nU.$$

Then R_n/nU is continuous. If $x \in nU$, then if $\lambda_0 T(x) \in nU$,

$$R_n(x) = \lambda_0 T(x) \in nU.$$

If $\lambda_0 T(x) \in \mathcal{N} - nU$, then

$$\|R_n(x)\| = \frac{n}{\|\lambda_0 T(x)\|} \lambda_0 T(x) = n.$$

Thus if $x \in nU$, then $R_n(x) \in nU$. Since T is compact, then R_n is compact. Since nU is a bounded convex closed set, then by the Schauder Fixed Point Theorem (4.1) there is an $x_n \in nU$ such that

$$R_n(x_n) = x_n.$$

Now suppose $\lambda_0 T(x_n) \in nU$. Then

$$\lambda_0 T(x_n) = R_n(x_n) = x_n,$$

which contradicts the property of λ_0. Therefore

$$\lambda_0 T(x_n) \in E - nU$$

and

$$R_n(x_n) = \frac{n}{\|\lambda_0 T(x_n)\|} \lambda_0 T(x_n) = x_n$$

and

$$\|x_n\| = n.$$

Also $\|\lambda_0 T(x_n)\| > n$. Then

$$\frac{n}{\|\lambda_0 T(x_n)\|} = \mu_n$$

where $0 < \mu_n < 1$. ∎

5. **Definition and properties of the Leray-Schauder degree.** We use the approximating transformations $T_n = F_{\varepsilon_n} T$ used in the proof of the Schauder Fixed Point Theorem (4.1) to define the Leray-Schauder degree for mappings of the form $I - T$ where I is the identity transformation and T is a compact transformation. Let ω be a bounded open set in $\mathcal{N}$, a normed linear space, and $\bar{\omega}$ its closure.

Now let $\omega' = \bar{\omega} - \omega$ be the boundary of ω and suppose $\bar{0} \notin (I - T)(\omega')$. We will define $d[I - T, \bar{\omega}, \bar{0}]$, i.e., the local degree of the mapping $I - T$ at $\bar{0}$ and relative to $\bar{\omega}$. (For convenience in notation and computation, we define the local degree at $\bar{0}$. It is routine to extend the definition to that of local degree at an arbitrary point in $\mathscr{N}$.) First we show that there is a positive number r such that

$$\operatorname*{glb}_{y \in \omega'} \|(I - T)(y) - \bar{0}\| \geqq r.$$

Suppose there is a sequence $\{y_m\}$ contained in ω' such that

$$\lim_{m \to \infty} (I - T)(y_m) = \bar{0}.$$

Since $[T(y_m)]$ is contained in the compact set $T(\bar{\omega})$, there exists y_0 such that $\{T(y_m)\}$ or a subsequence of it converges to y_0. For simplicity of notation, assume $\{T(y_m)\}$ converges to y_0.

Hence

$$\lim_{m \to \infty} y_m = \lim_{m \to \infty} [T(y_m)] = y_0.$$

But $[y_m] \subset \omega'$, a closed set. Therefore $y_0 \in \omega'$ and

$$(I - T)(y_0) = \bar{0}.$$

This contradicts the assumption that $\bar{0} \notin (I - T)(\omega')$.

Let $\{\varepsilon_n\}$ be a monotonic decreasing sequence of positive numbers with limit 0 and take integer n such that

$$\varepsilon_n < \frac{r}{2}.$$

Then for $x \in \omega'$,

$$\|(I - T_n)x - \bar{0}\| \geqq \frac{r}{2}.$$

Let $\mathscr{S}_n$ be a finite-dimensional subspace of $\mathscr{N}$ which contains $v_1, \cdots, v_{p_n}$, an ε_n-net for $\overline{T(\bar{\omega})}$, and at least one point of ω. The norm given on $\mathscr{N}$ is also a norm for the linear space $\mathscr{S}_n$. Hence $\mathscr{S}_n$ is a normed linear space and it is easy to show that

$$\mathscr{S}_n \cap \omega$$

is a nonempty bounded open set ω_n in $\mathscr{S}_n$ and $\omega_n' \subset \omega'$.

Since

$$(I - T_n)\bar{\omega}_n \subset \mathscr{S}_n$$

and

$$\operatorname*{glb}_{x \in \omega_n'} \|(I - T_n)(\omega_n') - \bar{0}\| \geqq \frac{r}{2},$$

then in the space $\mathscr{S}_n$, the local degree $d[I - T_n, \omega_n, \bar{0}]$ is defined.

Definition. The Leray-Schauder degree or LS degree of $I - T$ at $\bar{0}$ and relative to $\bar{\omega}$, denoted by $d[I - T, \bar{\omega}, \bar{0}]$, is $d[I - T_n, \bar{\omega}_n, \bar{0}]$.

To justify this definition, it must be shown that $d[I - T, \bar{\omega}, \bar{0}]$ is independent of the choice of T_n. Suppose T_{n_1} and T_{n_2} are two mappings such that $\varepsilon_{n_1} < r/2$ and $\varepsilon_{n_2} < r/2$. Let $\mathscr{S}_{n_1}$, $\mathscr{S}_{n_2}$ be the corresponding finite-dimensional subspaces. Let $\mathscr{S}_m$ be the finite-dimensional subspace spanned by the elements of $\mathscr{S}_{n_1}$ and $\mathscr{S}_{n_2}$. By the Reduction Theorem (10.1) of Chapter I,

$$d[I - T_{\varepsilon_{n_i}}, \bar{\omega}_{n_i}, \bar{0}] = d[I - T_{\varepsilon_{n_i}}, \bar{\omega}_m, \bar{0}] \quad (i = 1, 2) \tag{5.1}$$

where

$$\bar{\omega}_m = \mathscr{S}_m \cap \bar{\omega} \quad \text{and} \quad \bar{\omega}_{n_i} = \mathscr{S}_{n_i} \cap \bar{\omega}.$$

Consider the homotopy

$$H(x, t) = t(I - T_{\varepsilon_{n_1}})(x) + (1 - t)(I - T_{\varepsilon_{n_2}})(x)$$

with $t \in [0, 1]$. For all $x \in \omega'$,

$$\begin{aligned}\|t(I - T_{\varepsilon_{n_1}})(x) + (1 - t)(I - T_{\varepsilon_{n_2}})(x) - (I - T)(x)\| \\ \leqq \|t(I - T_{\varepsilon_{n_1}})(x) - t(I - T)(x)\| \\ + \|(1 - t)(I - T_{\varepsilon_{n_2}})(x) - (1 - t)(I - T)(x)\| \\ \leqq t\varepsilon_{n_1} + (1 - t)\varepsilon_{n_2} \leqq \frac{r}{2}.\end{aligned}$$

Hence for $x \in \mathscr{S}_m \cap \omega'$ and $t \in [0, 1]$,

$$\begin{aligned}\|H(x, t) - \bar{0}\| &= \|H(x, t) - (I - T)x + (I - T)x - \bar{0}\| \\ &\geqq \|(I - T)x - \bar{0}\| - \|H(x, t) - (I - T)x\| \\ &\geqq r - \frac{r}{2} = \frac{r}{2}.\end{aligned}$$

Then by Corollary (6.5) of Chapter I,

$$d[I - T_{\varepsilon_{n_1}}, \bar{\omega}_m, \bar{0}] = d[I - T_{\varepsilon_{n_2}}, \bar{\omega}_m, \bar{0}]. \tag{5.2}$$

Equations (5.1) and (5.2) give the desired result.

Now we show how the proofs that LS degree has the usual properties of local degree follow easily from the corresponding properties of local degree in Euclidean space.

Property 1: If $d[I - T, \bar{\omega}, \bar{0}] \neq 0$, then there is an $x \in \omega$ such that

$$(I - T)x = \bar{0}.$$

By the corresponding statement for finite-dimensional local degree (Theorem (6.6) of Chapter I), there is an $x_n \in \omega_n$ such that

$$(I - T_n)x_n = \bar{0}.$$

Hence

$$\|(I - T)x_n - \bar{0}\| < \varepsilon_n$$

or

$$(I - T)x_n = \eta_n \tag{5.3}$$

where

$$\|\eta_n\| < \varepsilon_n.$$

Since T is completely continuous and $\{x_n\} \subset \omega$, a bounded set, then $[T(x_n)]$ is a compact set and there is a subsequence $\{x_{n_m}\}$ of $\{x_n\}$ and a point y such that

$$\lim_{m\to\infty} T(x_{n_m}) = y. \tag{5.4}$$

From (5.3) and (5.4) and the fact that the sequence $\{\eta_n\}$ converges to $\bar{0}$, we obtain:

$$\lim_{m\to\infty} x_{n_m} = y.$$

Since $I - T$ is continuous, then

$$\lim_{m\to\infty} (I - T)x_{n_m} = (I - T)y.$$

But

$$(I - T)x_{n_m} = \eta_{n_m}.$$

Hence

$$(I - T)y = \bar{0}. \ \blacksquare$$

DEFINITION. Let $T(\tau)$ be a mapping from $[0, 1]$ into the set of compact transformations of a subset of normed linear space $\mathscr{N}$ into $\mathscr{N}$. That is, corresponding to each $\tau \in [0, 1]$, there is a compact transformation $T(\tau)$ of a subset E of $\mathscr{N}$ into $\mathscr{N}$. Then mapping T is a *homotopy of compact transformations on E* if: given $\varepsilon > 0$ and an arbitrary bounded set $M \subset E$, then there is a $\delta > 0$ such that if

$$|\tau_1 - \tau_2| < \delta$$

then for all $x \in M$

$$\|[T(\tau_1)]x - [T(\tau_2)]x\| < \varepsilon.$$

PROPERTY 2. Invariance under homotopy: if $T(\tau)$ is a homotopy of compact transformations on $\bar{\omega}$ where ω is a bounded open set, and if for all $x \in \omega'$ and all $\tau \in [0, 1]$,

$$(I - T(\tau))x \neq \bar{0},$$

then for all $\tau \in [0, 1]$, $d[I - T(\tau), \bar{\omega}, \bar{0}]$ exists and has the same value.

Proof. First we show that there is a positive number r such that if $x \in \omega'$, then for all $\tau \in [0, 1]$,

$$\|(I - T(\tau))x - \bar{0}\| \geqq r.$$

Suppose that for each positive integer n, there is an $x_n \in \omega'$ and $\tau_n \in [0, 1]$ such that

$$[I - T(\tau_n)]x_n - \bar{0} = \eta_n,$$

where

$$\|\eta_n\| < \frac{1}{n}.$$

Then

$$x_n = [T(\tau_n)]x_n + \eta_n. \tag{5.5}$$

Since $\{x_n\} \subset \omega'$, then $\{x_n\}$ is bounded. Since $\{\tau_n\} \subset [0, 1]$, there is a $\tau_0 \in [0, 1]$ such that a subsequence of $\{\tau_n\}$ converges to τ_0. For simplicity, we denote this subsequence by $\{\tau_n\}$. The point set corresponding to the sequence $\{[T(\tau_0)](x_n)\}$ is compact; hence there is a subsequence $\{[T(\tau_0)](x_{n_m})\}$ which converges to y where $y \in \omega'$. Since

$$\lim_{n \to \infty} \tau_n = \tau_0,$$

then

$$\lim_{m \to \infty} [T(\tau_{n_m})](x_{n_m}) = y.$$

Hence by (5.5),

$$\lim_{m \to \infty} x_{n_m} = y. \tag{5.6}$$

From equation (5.5) and the fact that

$$\lim_{m \to \infty} [T(\tau_0)](x_{n_m}) = y,$$

it follows that

$$[I - T(\tau_0)]y = \bar{0}.$$

Since $y \in \omega'$, this contradicts the hypothesis.

To prove that Property 2 holds, let $\tau_1 \in (0, 1)$. Let $T_n = F_{\varepsilon_n} T(\tau_1)$ be an approximating mapping from $\mathcal{N}$ into a finite-dimensional subspace $\mathcal{S}_n$ of the kind used in the definition of the LS degree such that for $x \in \mathcal{N}$,

$$\|T_n(x) - [T(\tau_1)](x)\| < \tfrac{1}{4}r.$$

Since $T(\tau)$ is a homotopy, there is a $\delta > 0$ such that if

$$|\tau - \tau_1| < \delta,$$

then for $x \in \omega'$,

$$\|[T(\tau)]x - [T(\tau_1)]x\| < \tfrac{1}{4}r.$$

Hence if $|\tau - \tau_1| < \delta$ and $x \in \omega'$

$$\|[T(\tau)]x - T_n(x)\| < \tfrac{1}{2}r.$$

By the definition of the LS degree, for τ such that $|\tau - \tau_1| < \delta$,

$$d[I - T(\tau), \bar{\omega}, \bar{0}] = d[I - T_n, \bar{\omega}_n, \bar{0}]$$

where $\bar{\omega}_n = \bar{\omega} \cap \mathscr{S}_n$. The result then follows from application of the Heine-Borel Theorem. ∎

6. **Proof of the Schauder Fixed Point Theorem by using the LS degree.** Just as in the finite-dimensional case, we may prove the fixed point theorem by using local degree.

Let T be a compact transformation of $\bar{\omega}$, a convex set which is the closure of a bounded open set, into itself. If there is a point $x \in \omega'$ such that $(I - T)x = \bar{0}$, the theorem is true. Suppose there is no $x \in \omega'$ such that $(I - T)x = \bar{0}$. Then $d[I - T, \bar{\omega}, \bar{0}]$ is defined. We consider the homotopy $I - tT$ where $t \in [0, 1]$. For simplicity, assume $\bar{\omega}$ is the sphere of unit radius and center $\bar{0}$. If there is an $x \in \omega'$ such that for some $t \in (0, 1)$

$$x = tT(x)$$

then $\|tT(x)\| = \|x\| = 1$. Therefore $\|T(x)\| > 1$ and T is not a mapping into $\bar{\omega}$. If there is an $x \in \omega'$ such that

$$x = tT(x)$$

for $t = 0$, we have: $x = \bar{0}$, a contradiction to the assumption that $x \in \omega'$. Since $d[I, \bar{\omega}, \bar{0}] = 1$, the Schauder Fixed Point Theorem follows from Property 2 and Property 1. ∎

7. **Computation of the LS degree and how the LS degree indicates the number of solutions.** One of the chief barriers to effective use of the LS degree is the fact that there are very few methods for computing the LS degree. For example, except for the case of computing the LS degree relative to small sets (which we study in § 8 of this chapter) the only LS degrees which have been computed have value $+1$.

In applications, the LS degree is usually determined by a simple use of the invariance under homotopy property as in the following example.

(7.1) DEFINITION. Let G be a transformation of $\mathscr{B}$ into itself. There exists *an a priori bound* for solutions of

$$G(x) = \bar{0} \tag{7.2}$$

if there is a positive number A such that if x_0 is a solution of $G(x) = \bar{0}$, then $\|x_0\| \leqq A$. The number A is called an *a priori bound*. (Note that the existence of an a priori bound says nothing about the number of solutions of equation (7.2) or if indeed the equation has any solutions at all.)

(7.3) THEOREM. *Suppose that $I - T(t)$ is a homotopy on $\mathcal{N}$ and suppose that for all $t \in [0, 1]$, there exists an a priori bound A for the solutions of*

$$[I - T(t)]x = \bar{0}$$

and the a priori bound is independent of t. Let

$$\bar{\omega} = [x \in \mathcal{B} / \|x\| \leqq rA]$$

where r is a constant greater than one. Then for all $t \in [0, 1]$, the LS degree

$$d[I - T(t), \bar{\omega}, \bar{0}]$$

is defined and has the same value.

PROOF. The proof follows directly from Property 2 (invariance under homotopy) of the LS degree.

Schaefer's Theorem (4.4) is a corollary of Theorem (7.3). If we take $T(t) = tT$ where T is a fixed compact transformation and observe that the local degree of the identity transformation is $+1$, we obtain Schaefer's Theorem from Theorem (7.3) by using Property 1 of the LS degree.

In Chapter IV, we will see that all the applications that are handled by applying Theorem (7.3) can also be dealt with by using Schaefer's Theorem because all the applications that have been studied are cases in which $I - T(t)$ has the form $I - tT$. Theoretically, however, the LS degree could be used to deal with a class of problems much wider than the class studied by use of Schaefer's Theorem.

Even if the LS degree could be computed, it would still give only very limited information about the number of solutions of the corresponding equation. There is no theorem analogous to Theorem (7.3) of Chapter I for the LS degree. The only known result in this direction is the following theorem whose proof we omit (see Cronin [**3**]).

THEOREM. *Suppose $d[I - T, \bar{\omega}, \bar{0}] = m$ where $|m| > 1$ and suppose $(I - T)(\bar{0}) = y$. Then if U is an arbitrary neighborhood of y, there is an element $y_1 \in U$ and a neighborhood V of $\bar{0}$ such that V contains at least two distinct points which map under $I - T$ into q.*

The proof of this theorem requires the fact (proved by Leray [**1**]) that if $I - T$ is a homeomorphism on $\bar{\omega}$ and if $q \in (I - T)\omega$, then the LS degree of $I - T$ at q is $+1$ or -1.

8. A partially analytic approach.

So far we have worked entirely in a normed linear space. Now we shall require that our space be complete, i.e., that we work in a Banach space. With this hypothesis we first obtain another fixed point theorem (the Banach Fixed Point Theorem). The mappings considered are contraction mappings; the theorem is proved by successive approximations; and the fixed point obtained is unique.

Then we will combine the approaches used in the Leray-Schauder theory and the Banach Fixed Point Theorem. For this case we will be able to obtain our most explicit results about the existence and number of solutions. However, these results will be local in that mappings on "small" sets will be considered.

DEFINITION. Let T be a continuous transformation from E, a subset of a Banach space $\mathscr{B}$, into $\mathscr{B}$.

Transformation T is a *contraction mapping* if there is a number $c \in (0, 1)$ such that if $x_1, x_2 \in E$, then

$$\|T(x_1) - T(x_2)\| \leqq c\|x_1 - x_2\|.$$

(8.1) BANACH FIXED POINT THEOREM. *If T is a contraction mapping with domain E, a closed subset of a Banach space $\mathscr{B}$, such that $T(E) \subset E$, then T has a unique fixed point in E.*

PROOF. Let $x_0 \in E$. If $T(x_0) = x_0$, then x_0 is a fixed point. If $T(x_0) \neq x_0$, define the sequence:

$$x_1 = T(x_0)$$

$$\cdots$$

$$x_{n+1} = T(x_n), \qquad n = 1, 2, \cdots.$$

(This is usually called a sequence of successive approximations.)

Since

$$\|x_{n+1} - x_n\| = \|T(x_n) - T(x_{n-1})\| \leqq c\|x_n - x_{n-1}\|,$$

then

$$\|x_{n+1} - x_n\| \leqq c^n\|x_1 - x_0\|.$$

Since $c < 1$, the sequence $\{x_n\}$ is a Cauchy sequence in $\mathscr{B}$ and hence has a limit $\bar{x}$. And $\bar{x} \in E$ because E is closed. Since T is continuous, then taking the limit in n on both sides of the equation

$$x_{n+1} = T(x_n),$$

we obtain

$$\bar{x} = T(\bar{x}),$$

i.e., $\bar{x}$ is a fixed point under T. The fixed point $\bar{x}$ is unique because suppose there are two fixed points $\bar{x}$ and $\bar{x}_1$. Then

$$\bar{x} = T(\bar{x})$$

and

$$\bar{x}_1 = T(\bar{x}_1).$$

Therefore

$$\|\bar{x} - \bar{x}_1\| = \|T(\bar{x}) - T(\bar{x}_1)\| \leqq c\|\bar{x} - \bar{x}_1\|.$$

Since $c < 1$, this implies: $\|\bar{x} - \bar{x}_1\| = 0$. ∎

Now we study a combination of the Leray-Schauder theory and the Banach Fixed Point Theorem. For this we shall need some further Banach space theory, in particular the theory of Riesz operators, i.e., operators of the form $I + C$ where C is linear and compact.

DEFINITION. A *linear functional* on Banach space $\mathscr{B}$ is a mapping from $\mathscr{B}$ into the real numbers such that:

(1) $f(a_1x_1 + a_2x_2) = a_1f(x_1) + a_2f(x_2)$;

(2) there is a positive number M such that for all $x \in \mathscr{B}$,

$$|f(x)| \leqq M\|x\|.$$

From condition (2), it follows that a linear functional is continuous. It is not difficult to prove that $\mathscr{B}^*$, the set of linear functionals, is itself a linear space; that

$$\|f\| = \operatorname*{lub}_{x \in \mathscr{B}} \frac{|f(x)|}{\|x\|}$$

is a norm for $\mathscr{B}^*$; and with this norm, $\mathscr{B}^*$ is a Banach space.

DEFINITION. The Banach space $\mathscr{B}^*$ is the *conjugate space of* $\mathscr{B}$.

Now suppose T is a linear transformation of $\mathscr{B}$ into $\mathscr{B}$. We define a transformation T^* of $\mathscr{B}^*$ into $\mathscr{B}^*$ in this way: if f is a linear functional on $\mathscr{B}$, then for each $x \in \mathscr{B}$, define:

$$[T^*(f)](x) = f[T(x)].$$

The $T^*(f)$ thus defined in a linear functional and the transformation T^* is a linear transformation $\mathscr{B}^*$ into $\mathscr{B}^*$.

DEFINITION. T^* is the *conjugate transformation* of T.

Let C be a linear compact transformation of $\mathscr{B}$ into $\mathscr{B}$. Now we summarize those parts of the theory of transformations of the form $I + C$ (Riesz transformations) which we will need here. (See Banach [**1**], Bers [**1**], Riesz [**1**], Riesz and Sz.-Nagy [**1**], and Bartle [**1**].)

The set

$$\mathscr{B}_n = [x \in \mathscr{B} \mid (I + C)^n x = 0] \qquad (n = 1, 2, \cdots)$$

is a linear finite-dimensional space. If $\mathscr{B}_1 = \bar{0}$, then transformation $I + C$ has an inverse on $\mathscr{B}$ and the inverse is a linear transformation. If $\mathscr{B}_1 \neq \bar{0}$, there is a positive integer m such that if $n < m$, then

$$\mathscr{B}_n \subsetneqq \mathscr{B}_{n+1}$$

and if $n \geqq m$, then

$$\mathscr{B}_n = \mathscr{B}_{n+1}.$$

The number m is called the *Riesz index* of transformation $I + C$. Now let

$$\mathscr{B}^{(n)} = (I + C)^n \mathscr{B}.$$

For each n, the set $\mathscr{B}^{(n)}$ is a subspace of $\mathscr{B}$. If $n < m$, then

$$\mathscr{B}^{(n)} \supsetneqq \mathscr{B}^{(n+1)}.$$

If $n \geqq m$, then

$$\mathscr{B}^{(n)} = \mathscr{B}^{(n+1)}.$$

Using these facts, it can be proved that $\mathscr{B}_1 \cap \mathscr{B}^{(1)} = \bar{0}$ and $\mathscr{B}$ is the direct sum of $\mathscr{B}_1$ and $\mathscr{B}^{(1)}$, i.e., if $x \in \mathscr{B}$, then there is a unique $x_1 \in \mathscr{B}_1$ and a unique $x^1 \in \mathscr{B}^{(1)}$ such that

$$x = x_1 + x^1.$$

Let E_1, E^1 be the transformations defined by:

$$E_1 : x \to x_1,$$

$$E^1 : x \to x^1.$$

Transformations E_1 and E^1 are linear transformations of $\mathscr{B}$ into $\mathscr{B}$. The transformations E_1E^1 and E^1E_1 are both the null transformation (i.e., the transformation which takes every element of $\mathscr{B}$ into $\bar{0}$) and

$$(E_1)^2 = E_1$$

and

$$(E^1)^2 = E^1.$$

The conjugate transformation of $I + C$ is $I^* + C^*$ where I^* is the identity transformation on $\mathscr{B}^*$ and C^* is the conjugate transformation of C. It follows that C^* is compact and if

$$\mathscr{B}_1^* = [f \in \mathscr{B}^* \,/\, (I^* + C^*)f = \bar{0}],$$

then $\mathscr{B}_1^*$ is finite-dimensional and the dimension of $\mathscr{B}_1^*$ is equal to the dimension of $\mathscr{B}_1$.

Let $x_1, \cdots, x_n$ be a basis of $\mathscr{B}_1$ and let $f_1, \cdots, f_n$ be a basis of $\mathscr{B}_1^*$. Then there exist $g_1, \cdots, g_n \in \mathscr{B}^*$ such that for $x \in \mathscr{B}$,

$$E_1(x) = \sum g_i(x)x_i$$

(then $g_i(x_j) = \delta_{ij}$) and there exist $y_1, \cdots, y_n \in \mathscr{B}$ such that

$$f_i(y_j) = \delta_{ij}.$$

Now for all $x \in \mathscr{B}$, define the transformation $I + C_0$ by:

$$(I + C_0)x = (I + C)x + \sum_{i=1}^{n} g_i(x)y_i.$$

The transformation C_0 is compact and the set

$$\mathscr{B}_1^{(0)} = [x \in \mathscr{B} \,/\, (I + C_0)x = \bar{0}]$$

contains only the point $\bar{0}$. Hence $I + C_0$ has an inverse on $\mathscr{B}$ and the inverse is a linear transformation We denote this inverse by R.

(8.2) THEOREM. *For all* $x \in \mathscr{B}$,

$$R(1 + C)x = (I - E_1)x = x - \sum_{i=1}^{n} g_i(x)x_i$$

and

$$(I + C)Rx = x - \sum_{i=1}^{n} f_i(x)y_i.$$

PROOF.

$$R(I + C)x = R(I + C_0)x - R\sum_{i=1}^{n} g_i(x)y_i$$

$$= x - (I + C_0)^{-1} \sum_{i=1}^{n} g_i(x)y_i$$

$$= x - \sum_{i=1}^{n} g_i(x)(I + C_0)^{-1}y_i.$$

But

$$y_i = (I + C)x_i + \sum_{j=1}^{n} g_j(x_i)y_j = (I + C_0)x_i$$

or

$$(I + C_0)^{-1}y_i = x_i.$$

Therefore

$$R(1 + C)x = x - \sum_{i=1}^{n} g_i(x)x_i,$$

$$(I + C)Rx = (I + C_0)Rx - \sum_{i=1}^{n} g_i(Rx)y_i$$

$$= x - \sum_{i=1}^{n} g_i(Rx)y_i.$$

To complete the proof, we show: for all $x \in \mathscr{B}$,

$$g_i(Rx) = f_i(x);$$

or letting $Rx = y$,

$$g_i(y) = f_i(R^{-1}y).$$

But

$$f_i(R^{-1}y) = f_i[(I + C_0)y]$$

$$= f_i\left[(I + C)y + \sum_{i=1}^{n} g_i(y)y_i\right]$$

$$= g_i(y). \blacksquare$$

(8.3) THEOREM. *If $x \in \mathscr{B}$, then*

$$E_1 Rx = R\left[\sum_{i=1}^{n} f_i(x) y_i\right].$$

PROOF. From the definition of g_i,

$$\begin{aligned} E_1[R(x)] &= \sum_{i=1}^{n} g_i[R(x)]x_i \\ &= \sum_{i=1}^{n} g_i[R(x)]\{(I + C_0)^{-1}y_i\} \\ &= (I + C_0)^{-1} \sum_{i=1}^{n} \{g_i[R(x)]\}y_i \\ &= (I + C_0)^{-1} \sum_{i=1}^{n} [f_i(x)]y_i \\ &= R\left(\sum_{i=1}^{n} [f_i(x)]y_i\right). \blacksquare \end{aligned}$$

Finally we will need an implicit function theorem for Banach spaces.

(8.4) IMPLICIT FUNCTION THEOREM. *Let $\mathscr{X}$, $\mathscr{Y}$, $\mathscr{Z}$ be Banach spaces and U, V, W be open sets in $\mathscr{X}$, $\mathscr{Y}$, $\mathscr{Z}$, respectively. Let L be a function with domain $U \times V \times W$ and range a subset of $\mathscr{X}$. Assume the following conditions are satisfied:*

(H_1) *there is a point $(x_0, y_0, z_0) \in U \times V \times W$ such that*

$$x_0 = L(x_0, y_0, z_0);$$

(H_2) *there exists a positive number $c < 1$ such that*

$$\|L(x_1, y, z) - L(x_2, y, z)\| \leqq c\|x_1 - x_2\|$$

for every (x_1, y, z), (x_2, y, z) in $U \times V \times W$;

(H_3) *L is uniformly continuous on $U \times V \times W$.*

Then the following conclusions hold:

(C_1) *for each $(y, z) \in V \times W$, there is at most one point (x, y, z) in $U \times V \times W$ which is a solution of the equation:*

$$\text{(E)} \qquad x = L(x, y, z);$$

(C_2) *there exist open sets V_1 and W_1, spherical neighborhoods of y_0 and z_0 respectively, and a function F with domain $V_1 \times W_1$ and range a subset of $\mathscr{X}$ such that the point $(F(y, z), y, z)$ is an element of $U \times V \times W$ and is a solution of equation* (E) *for every $(y, z) \in V_1 \times W_1$;*

(C_3) *the solution $F(y, z)$ of equation* (E) *is uniformly continuous on $V_1 \times W_1$.*

This theorem was proved by Hildebrandt and Graves [**1**].

Now we are ready to study the solutions x of the equation in Banach space $\mathscr{B}$:

$$(I + C)x + T(x) = y \tag{8.5}$$

where y is given, transformation C is linear and compact, and transformation T satisfies the conditions:

(i) the domain of T is a sphere K in $\mathscr{B}$ with center at the origin;

(ii) $T(\bar{0}) = \bar{0}$;

(iii) if $x_a, x_b \in K$, then

$$\|T(x_a) - T(x_b)\| \leqq M(x_a, x_b)\|x_a - x_b\|$$

where $M(x_a, x_b)$ is a positive valued function such that

$$\lim_{(x_a, x_b) \to (\bar{0}, \bar{0})} M(x_a, x_b) = 0.$$

Since the theory to be described is an abstract version of part of the work of Erhard Schmidt [**1**] on nonlinear integral equations we will call an operator of the form $I + C + T$ a Schmidt operator. We note first that if the inverse $(I + C)^{-1}$ exists, then if equation (8.5) is multiplied by $(I + C)^{-1}$, the resulting equation can be solved locally for x, uniquely in terms of y, by applying the Implicit Function Theorem (8.4). We will be concerned with the case: $I + C$ does not have an inverse, i.e., $\mathscr{B}_1 \neq \bar{0}$.

Multiplying (8.5) by R and applying the first statement of Theorem (8.2) we obtain, denoting $E_1(x)$ and $E^1(x)$ by x_1 and x^1, respectively:

$$x - x_1 + RT(x_1 + x^1) = Ry$$

or

$$x^1 + RT(x_1 + x^1) = Ry. \tag{8.6}$$

Applying E_1 and E^1 to equation (8.6) we obtain the equations:

$$E_1RT(x_1 + x^1) = E_1R(y), \tag{8.7}$$

$$x^1 + E^1RT(x_1 + x^1) = E^1Ry. \tag{8.8}$$

(8.9) LEMMA. *There exist neighborhoods U, V, W of $x^1 = \bar{0}$, $x_1 = \bar{0}$ and $y = \bar{0}$ in $\mathscr{B}^{(1)}$, $\mathscr{B}_1$, and $\mathscr{B}$, respectively, such that the function*

$$-E^1RT(x_1 + x^1) + E^1Ry$$

may be regarded as a mapping from $U \times V \times W$ into $\mathscr{B}^{(1)}$ and as such satisfies the conditions (H_1), (H_2), *and* (H_3) *of Theorem* (8.4).

PROOF. Since $\mathscr{B}^{(1)}$ is closed and $\mathscr{B}_1$ is finite-dimensional, both are Banach spaces. That conditions (H_1), (H_2), and (H_3) are satisfied follows from the hypotheses on transformations C and T. ▮

Thus for sufficiently small x_1, x^1, and y, equation (8.8) can be solved for x^1 as a continuous function of x_1 and y, i.e.,

$$x^1 = x^1(x_1, y). \tag{8.10}$$

Substituting from (8.10) into (8.7), we obtain:

$$E_1 RT[x_1 + x^1(x_1, y)] - E_1 R(y) = \bar{0}. \tag{8.11}$$

Solving (8.5) for x as a function of y is clearly equivalent to solving (8.11) for x_1 as a function of y. We study the solutions of (8.11), which is an equation in Euclidean n-space (where n is the dimension of $\mathscr{B}_1$), by using local degree.

Now we sharpen the higher order condition on T by imposing the following hypothesis:

(iv) transformation T is a sum,

$$T(x) = T^{(k)}(x) + T^{(k+1)}(x),$$

where $T^{(k)}$ is a continuous mapping of the domain K of T in $\mathscr{B}$ and $T^{(k)}(x)$ is homogeneous of degree k in x; that is, if m is an arbitrary integer,

$$T^{(k)}\left(\sum_{i=1}^{m} a_i w_i\right) = \sum a_1^{\beta_1} \cdots a_m^{\beta_m} \, T_{\beta_1 \cdots \beta_m}(w_1, \cdots, w_m)$$

where the summation is taken over all integers β_i such that $\beta_i \geqq 0$ and $\sum_{i=1}^{m} \beta_i = k$ and $T_{\beta_1 \cdots \beta_m}(w_1, \cdots, w_m)$ is a continuous mapping from $K \times \cdots \times K$ into $\mathscr{B}$ which is independent of w_i if $\beta_i = 0$; and $T^{(k+1)}$ is a continuous mapping from K into $\mathscr{B}$ such that

$$\lim_{x \to \bar{0}} \frac{T^{(k+1)}(x)}{\|x\|^k} = \bar{0}.$$

Hypothesis (iv) implies the following condition which is sometimes easier to use:

(iv′) there is a maximal integer $k \geqq 2$ such that

$$T(ax) = a^k T^1(a, x)$$

where T^1 is a continuous mapping of an open subset of $R \times \mathscr{B}$ into $\mathscr{B}$ where R denotes the real numbers. This subset contains points of the form $(0, x)$ where 0 is the zero of R.

In order to be able to compute the local degree we impose also the condition: if $x_1 \in \mathscr{B}_1$ and $x_1 \neq \bar{0}$, then

$$E_1 RT^{(k)}(x_1) \neq \bar{0}. \tag{8.12}$$

(Theorem (8.3) is useful in verifying (8.12) in particular applications.) We also need more detailed information about the form of $x^1(x_1, y)$.

(8.13) LEMMA. *There are neighborhoods* $N(0)$ *and* $N(\bar{0})$ *of* $0 \in R$ *and* $\bar{0} \in \mathscr{B}$, *such that if* $x_1 = av_1$ *where* $a \in N(0)$ *and* $V_1 \in N(\bar{0})$, *then*

$$x^1(x_1, \bar{0}) = a^k G(a, v_1)$$

where $G(a, v_1)$ *is a continuous mapping from* $N(0) \times N(\bar{0})$ *into* $\mathscr{B}^{(1)}$ *and* k *is the integer described above in hypothesis* (iv) *on transformation* T.

PROOF. $x^1(x_1, \bar{0})$ is the solution of

$$x^1 + E^1RT(x_1 + x^1) = \bar{0}, \tag{8.14}$$

i.e., equation (8.8) with y set equal to $\bar{0}$. Let $x_1 = av_1$ and denote E^1RT by J:

$$x^1 + J(av_1 + x^1) = \bar{0}.$$

Writing $z = x^1/a$ and letting E^1RT^1 be denoted by J^1, we have:

$$az + J(av_1 + az) = \bar{0}$$

or

$$az + a^kJ^1(a, v_1 + z) = \bar{0}$$

or

$$z = -\, a^{k-1}J^1(a, v_1 + z). \tag{8.15}$$

From the hypotheses (ii) and (iii) on T, it follows that the Implicit Function Theorem (8.4) may be used to solve (8.15) for z in terms of a and v_1 in neighborhoods $N(0)$ and $N(\bar{0})$ in R and $\mathscr{B}$ respectively: $z = A(a, v_1)$. Substituting in (8.15), we obtain:

$$z = -\, a^{k-1}J^1(a, v_1 + A(a, v_1)),$$

$$x^1 = -\, a^kJ^1(a, v_1 + A(a, v_1))$$

which completes the proof. ∎

Now let M_y be the mapping of $\mathscr{B}_1$ into $\mathscr{B}_1$ defined by

$$M_y : x_1 \rightarrow E_1RT[x_1 + x^1(x_1, y)] - E_1Ry$$

and let $M^{(k)}$ and $\underline{M}^{(k)}$ be the mappings of $\mathscr{B}_1$ into $\mathscr{B}_1$ defined by

$$M^{(k)} : x_1 \rightarrow E_1RT^{(k)}[x_1 + x^1(x_1, \bar{0})]$$

and

$$\underline{M}^{(k)} : x_1 \rightarrow E_1RT^{(k)}[x_1].$$

By inequality (8.12), the local degree $d[\underline{M}^{(k)}, \bar{\sigma}, \bar{0}]$ is defined where $\bar{\sigma}$ is a sphere in $\mathscr{B}$ with center $\bar{0}$. If $\bar{\sigma}_1$ is a sphere such that

$$\bar{\sigma}_1 \subset N(\bar{0}),$$

where $N(\bar{0})$ is the neighborhood described in Lemma (8.13), then by inequality (8.12), there is a positive constant m such that if $v_1 \in \sigma_1'$, then

$$\|E_1RT^{(k)}(v_1)\| \geq m.$$

But if a is sufficiently small, then

$$\|E_1RT^{(k)}[v_1 + a^{k-1}G(a, v_1)] - E_1RT^{(k)}[v_1]\| < \tfrac{1}{2}m.$$

Hence by Corollary (6.5) of Chapter I,

$$d[M^{(k)}, \bar{\sigma}, \bar{0}] = d[\underline{M}^{(k)}, \bar{\sigma}, \bar{0}]$$

where $\bar{\sigma}$ is a sufficiently small sphere with center $\bar{0}$.

Next,

$$\text{(8.16)} \quad \begin{aligned} &\|E_1RT^{(k)}[x_1 + x^1(x_1, \bar{0})] - E_1RT[x_1 + x^1(x_1, \bar{0})]\| \\ &\qquad = \|E_1RT^{(k+1)}[x_1 + x^1(x_1, \bar{0})]\|. \end{aligned}$$

Let $x_1 = av_1$ where $a \in N(0)$ and $v_1 \in N(\bar{0})$, and $N(0)$ and $N(\bar{0})$ are the neighborhoods described in Lemma (8.13). If ε is a given positive number and if a is sufficiently small, then by hypothesis (iv) on the transformation T and by Lemma (8.13),

$$\text{(8.17)} \quad \begin{aligned} \|E_1RT^{(k+1)}[x_1 + x^1(x_1, \bar{0})]\| &= \|E_1RT^{(k+1)}[av_1 + a^kG(a, v_1)]\| \\ &\leqq |a|^k\varepsilon. \end{aligned}$$

But

$$E_1RT^{(k)}[av_1 + a^kG(a, v_1)] = a^kE_1RT^{(k)}[v_1 + a^{k-1}G(a, v_1)].$$

By inequality (8.12), if $v_1 \in \sigma_1'$ where $\bar{\sigma}_1$ is a fixed sphere with center $\bar{0}$ such that $\bar{\sigma}_1 \subset N(\bar{0})$, the neighborhood described in Lemma (8.13), and if a is sufficiently small, then there exists a positive number ε_0 such that

$$\text{(8.18)} \quad \|E_1RT^{(k)}[v_1 + a^{k-1}G(a, v_1)]\| > \varepsilon_0.$$

If the ε in (8.17) and a are sufficiently small, then

$$|a|^k\varepsilon < \varepsilon_0.$$

From (8.16), (8.17), (8.18), we obtain then by Corollary (6.5) of Chapter I: if $\bar{\sigma}$ has sufficiently small radius, then $d[M_0, \bar{\sigma}, \bar{0}]$ exists and

$$d[M_0, \bar{\sigma}, \bar{0}] = d[M_k, \bar{\sigma}, \bar{0}].$$

Since $x^1(x, y)$ is uniformly continuous in y, then if y is sufficiently small, it follows, again by Corollary (6.5) of Chapter I, that

$$d[M_y, \bar{\sigma}, \bar{0}] = d[M_0, \bar{\sigma}, \bar{0}].$$

Thus to study the solutions of (8.5) which are close to $\bar{0}$ for y sufficiently close to $\bar{0}$, we study $d[\underline{M}^{(k)}, \bar{\sigma}, \bar{0}]$. (The mapping $\underline{M}^{(k)}$ is described by n polynomials homogeneous of degree k in n variables where n is the dimension of $\mathscr{B}_1$.) This $d[\underline{M}^{(k)}, \bar{\sigma}, \bar{0}]$ yields a lower bound for the number of distinct solutions of (8.5) provided we impose a differentiability condition on transformation T.

(v) At each point x in some neighborhood of $\overline{0}$, transformation T has a differential $L_x(\Delta x)$ which satisfies a Lipschitz condition in x. That is, there is a neighborhood N_d such that for each $x \in N_d$ there is a continuous linear transformation L_x and a transformation Q_x both taking $\mathscr{B}$ into $\mathscr{B}$ and such that:

(1) for each $\Delta x \in \mathscr{B}$, $T(x + \Delta x) - T(x) = L_x(\Delta x) + Q_x(\Delta x)$;

(2) there is a positive constant K such that if $u, v \in N_d$, then

$$\|L_u - L_v\| \leqq K\|u - v\|;$$

(3) $\lim_{\Delta x \to \overline{0}} Q_x(\Delta x)/\|\Delta x\| = \overline{0}$.

DEFINITION. Suppose that $N \subset \mathscr{B}$ is the direct sum of N_1 and N^1 (written $N = N_1 \oplus N^1$) neighborhoods of $\overline{0}$ in $\mathscr{B}_1$ and $\mathscr{B}^1$, i.e.,

$$N = [x \mid x = x_1 + x^1 \text{ with } x_1 \in N_1 \text{ and } x^1 \in N^1].$$

Suppose that corresponding to each point $x_\nu^1 \in N^1$, there is a subset E_ν of N_1 of n-dimensional measure zero. A set $A \subset N$ is said to contain a *good many of the points of* N if

$$A \supset \bigcup_\nu [x_\nu^1 + (N_1 - E_\nu)].$$

We omit the proof of the following theorem (see Cronin [**2**]).

(8.19) THEOREM. *If* $d[\underline{M}^{(k)}, \bar{\sigma}, \overline{0}] \neq 0$, *there exist neighborhoods* N_x *and* N_y *of* $\overline{0}$ *in* $\mathscr{B}$ *such that for a good many of the points* $y \in N_y$, *the equation*

$$(I + C + T)x = y$$

has at least $|d[\underline{M}^{(k)}, \bar{\sigma}, \overline{0}]|$ *distinct solutions in* N_x.

The following theorem (see Cronin [**3**]) is an example of a case in which the value of $d[\underline{M}^{(k)}, \bar{\sigma}, \overline{0}]$ can be studied:

(8.20) THEOREM. *If the dimension of* $\mathscr{B}_1$ *is one, if transformation* T *satisfies conditions* (i) *through* (iv), *if condition* (8.12) *is satisfied and if* k *is odd, then*

$$d[\underline{M}^{(k)}, \bar{\sigma}, \overline{0}] \neq 0.$$

As might be expected, if T is completely continuous so that the Leray-Schauder degree of $I + C + T$ is defined, then the Leray-Schauder degree equals $d[\underline{M}^{(k)}, \bar{\sigma}, \overline{0}]$ or $-d[\underline{M}^{(k)}, \bar{\sigma}, \overline{0}]$. Since this fact is not of interest for the applications to analysis in the next chapter, we omit the proof which is given in Cronin [**1**].

CHAPTER IV

Applications to Integral Equations, Partial Differential Equations, and Ordinary Differential Equations with Large Nonlinearities

1. **Introduction.** Most of the applications of Chapter III will be to functional equations: integral equations and partial differential equations. In a sense, we use the infinite-dimensional theory of Chapter III in the study of ordinary differential equations because throughout Chapter II we used Theorem (1.3) (the Basic Existence Theorem) which is proved by successive approximations and hence may be regarded as an application of the Banach Fixed Point Theorem (Theorem (8.1), Chapter III). However, after application of the Basic Existence Theorem, the questions we studied in Chapter II became finite-dimensional.

We might expect this correspondence:

ordinary differential equations
  $\leftrightarrow$ finite-dimensional methods (plus successive approximations)

functional equations
  $\leftrightarrow$ infinite-dimensional methods

because boundary conditions for ordinary differential equations are specified by points in Euclidean space while boundary conditions for functional equations, e.g., the Dirichlet problem for elliptic equations, are given by functions. However, this correspondence provides us with only a very rough rule for how topological techniques are applied. The study of the Schmidt operator in § 8 of Chapter III, which will be applied here to elliptic differential equations, is carried out by using successive approximations and then a finite-dimensional method. Another exception is the application of the Cesari method [**1**; **4**; **5**] to the theory of periodic solutions of ordinary differential equations with large nonlinearities which will be described in this chapter. The two main steps in the Cesari method are first the application of the Schauder Fixed Point Theorem (Theorem (4.1), Chapter III) or, in certain cases, the Banach Fixed Point Theorem (Theorem (8.1), Chapter III) and secondly a finite-dimensional study. These seem to be genuine exceptions to the correspondence described above. Infinite-dimensional methods have also been used in problems in ordinary differential equations in cases where successive approximations would suffice. For example the results of Kyner [**1**; **2**] obtained by use of the Schauder Fixed Point Theorem are contained in the later work of Hale [**1**] who uses contraction mappings and, thus, successive approximations. Also Mikolajska [**1**] has used the Schauder

Fixed Point Theorem to obtain results for ordinary differential equations which could be obtained by successive approximations.

In Chapter II, we were concerned with finding solutions with particular properties: periodicity, almost periodicity, stability. But here we will derive only existence theorems. The finer question of how many solutions exist can be treated only in applications of the theory of the Schmidt operator and in the Cesari theory; and we do not consider stability questions at all. That we are concerned primarily with existence theorems is not surprising because in the theory of nonlinear integral equations and nonlinear partial differential equations there is no general existence theory comparable to the theory for ordinary differential equations which is contained in the Basic Existence Theorem (1.3) and the Continuation Theorem (1.7) of § 1 of Chapter II. Indeed, present knowledge of nonlinear partial differential equations is somewhat fragmentary.

The topological study of functional equations requires far more extensive theory from analysis than the study of ordinary differential equations. In Chapter II, we were able to give a fairly complete account of the requisite theory for ordinary differential equations. But in order to keep Chapter IV of reasonable length, we shall be forced merely to quote the results about partial differential equations which will be used. Also the account will be less complete than for the ordinary differential equations because we will describe only some typical applications.

2. **Integral equations.** An equation of the form

$$\phi(s) + \int_a^b K[s, t, \phi(t)]\, dt = f(s), \tag{2.1}$$

where $f(s)$ and $K(s, t, \omega)$ are given and $\phi(s)$ is to be determined, is called a Fredholm integral equation of the second kind. Besides being of intrinsic value, this kind of equation often arises in the study of partial differential equations as we shall see later (see § 9 of this chapter). We will show how the Banach and Schauder Fixed Point Theorems (Theorems (4.1) and (8.1) of Chapter III) can be applied to equation (2.1).

Fredholm equations of the first kind, i.e., of the form

$$\int_a^b K[s, t, \phi(t)]\, dt = f(s),$$

where $f(s)$ and $K(s, t, \omega)$ are given and $\phi(s)$ is to be determined, must be treated differently. A primary source of this difference and the reason why there is not as "nice" a theory for equations of the first kind as for equations of the second kind is this: if function K is "well behaved," then function f is, in general, better behaved than ϕ. E.g., if $f(s)$ is differentiable, then ϕ may be continuous but not differentiable. See Courant and Hilbert [**1**, pp. 159–160].

We assume that $K(s, t, \omega)$ is continuous in $[a, b] \times [a, b] \times R$ where R is the space of real numbers. Let C denote the Banach space of real-valued continuous functions $g(s)$ having domain $[a, b]$ and with identity element $\bar{0}$ and

$$\|g\| = \max_{s \in [a, b]} |g(s)|.$$

Then the transformation $\mathscr{K}$ which takes $\phi(s)$ into

$$\int_a^b K[s, t, \phi(t)]\, dt$$

is a transformation from C into itself. Let $\phi_1, \phi_2 \in C$. Since K is continuous in ω for all R, then K is uniformly continuous in ω on any closed bounded set in R. In particular, K is uniformly continuous on $[a, b] \times [a, b] \times [-M, M]$ where M is the larger of the numbers $\|\phi_1\|$ and $\|\phi_2\|$. Hence if

$$\|\phi_1 - \phi_2\| < \delta$$

then for all $(s, t) \in [a, b] \times [a, b]$,

$$|K[s, t, \phi_1(t)] - K[s, t, \phi_2(t)]| < \varepsilon,$$

and

$$\begin{aligned} &\left| \int_a^b K[s, t, \phi_1(t)]\, dt - \int_a^b K[s, t, \phi_2(t)]\, dt \right| \\ &\qquad = \left| \int_a^b \{K[s, t, \phi_1(t)] - K[s, t, \phi_2(t)]\}\, dt \right| < \varepsilon(b - a). \end{aligned} \tag{2.2}$$

Thus transformation $\mathscr{K}$ is a continuous mapping from C into itself. Now assume that for all $(s, t) \in [a, b] \times [a, b]$ and all pairs $\omega_1, \omega_2 \in R$,

$$|K(s, t, \omega_1) - K(s, t, \omega_2)| < c|\omega_1 - \omega_2| \tag{2.3}$$

where c is a positive constant such that

$$c < \frac{1 - \varepsilon_0}{b - a}$$

where $0 < \varepsilon_0 < 1$. Then

$$\|\mathscr{K}(\phi_1) - \mathscr{K}(\phi_2)\| < (1 - \varepsilon_0)\|\phi_1 - \phi_2\|. \tag{2.4}$$

Thus if f is a fixed element of C, the transformation

$$\mathscr{T}(\phi) = f - \mathscr{K}(\phi)$$

is a contraction mapping. Now suppose $\mathscr{K}(\bar{0}) = \bar{0}$ and let

$$S = [g \in C \,/\, \|g\| < r].$$

Then if $g \in S$,

$$\begin{aligned}\|\mathcal{T}(g) - f\| &= \|f - \mathcal{K}(g) - f\| \\ &= \|\mathcal{K}(g)\| = \|\mathcal{K}(g) - \mathcal{K}(\bar{0})\| \\ &\leqq (1 - \varepsilon_0)\|g\| < (1 - \varepsilon_0)r.\end{aligned}$$

But

$$\|\mathcal{T}(g)\| \leqq \|\mathcal{T}(g) - f\| + \|f\|.$$

Thus if $\|f\| < \varepsilon_0 r$, then $\mathcal{T}(S) \subset S$. The hypotheses of the Banach Fixed Point Theorem (Theorem (8.1) of Chapter III) are satisfied and equation (2.1) has a unique solution ϕ in S. Note that we use the continuity of K and condition (2.3) only on $[a, b] \times [a, b] \times [-r, r]$. Examples of integral operators which satisfy (2.3) and the condition $K(\bar{0}) = \bar{0}$ are often obtained by requiring that K satisfy a "higher order condition" on ω. For example, let

$$K(s, t, \omega) = U(s, t)\omega^2$$

where $U(s, t)$ is a continuous function of (s, t) on $[a, b] \times [a, b]$. Let S be a sphere of radius r and center 0 in C where

$$r < (1 - \varepsilon_0)[(2 \max |U(s, t)|)(b - a)]^{-1},$$

where the maximum is taken for $(s, t) \in [a, b] \times [a, b]$. Then if $\phi_1, \phi_2 \in S$

$$\begin{aligned}|K(s, t, \phi_1) - K(s, t, \phi_2)| &\leqq |U(s, t)|\,|\phi_1 + \phi_2|\,|\phi_1 - \phi_2| \\ &\leqq |U(s, t)|\{|\phi_1| + |\phi_2|\}|\phi_1 - \phi_2| \\ &\leqq |U(s, t)|(2r)|\phi_1 - \phi_2| \\ &\leqq \frac{1 - \varepsilon_0}{(b - a)}|\phi_1 - \phi_2|.\end{aligned}$$

Similarly for any integer $m > 1$,

$$K(s, t, \omega) = U(s, t)\omega^m$$

can be shown to give rise to an operator $\mathcal{K}$ satisfying the required conditions.

Now we suppose that operator $\mathcal{K}$ does not satisfy a condition of the form (2.3). Generally we will be able to apply the Schauder Fixed Point Theorem (Theorem 4.1 of Chapter III) even if $K(\bar{0}) \neq \bar{0}$ because the operator K is compact. We show now that this compactness is a simple consequence of the following classical theorem in analysis:

(2.5) THEOREM (ASCOLI'S THEOREM). *If $\{f_\nu\}$ is a uniformly bounded equicontinuous set of real-valued functions defined on a compact (i.e., closed and bounded) subset A of R^1, then there is a sequence $\{f_n\}$ such that $\{f_n\} \subset \{f_\nu\}$ and $\{f_n\}$ converges uniformly on A to a (continuous) function f.*

PROOF. See Rudin [**1**, pp. 127 ff.] or McShane and Botts [**1**, pp. 92 ff.].

In order to prove that $\mathscr{K}$ is compact we need only show that if $\{\phi_\nu\}$ is a bounded set in C, then $\{\mathscr{K}(\phi_\nu)\}$ contains a uniformly convergent subsequence. From Theorem (2.5), we see that it is sufficient to show that the functions $\{\mathscr{K}(\phi_\nu)\}$ are uniformly bounded and equicontinuous. The uniform boundedness follows from the fact that $\{\phi_\nu\}$ is a bounded set in C, i.e., there is a positive number M such that $\|\phi_\nu\| < M$ for all ν, and that K is bounded on any set of the form $[a, b] \times [a, b] \times [-r, r]$ where r is a fixed positive number. The equicontinuity is proved as follows: first

$$|(\mathscr{K}\phi_n)(s_1) - (\mathscr{K}\phi_n)(s_2)| = \left| \int_a^b \{K[s_1, t, \phi_n(t)] - K[s_2, t, \phi_n(t)]\}\, dt \right| .$$

Since K is continuous as a function of s uniformly in (t, ω) for $(t, \omega) \in [a, b] \times [-M, M]$, if

$$|s_1 - s_2| < \delta,$$

then

$$|(K\phi_n)(s_1) - (K\phi_n)(s_2)| < \varepsilon(b - a).$$

Since the subscript n plays no role in the argument, the desired equicontinuity is obtained.

Thus $\mathscr{K}$ is compact and if $\mathscr{K}$ is a mapping from a bounded convex set into itself, we may apply the Schauder Fixed Point Theorem. As an example of a function K which gives rise to an operator $\mathscr{K}$ with the desired properties but which is not a contraction mapping, let

$$K(s, t, \omega) = U(s, t)(\cos \omega)^m$$

where m is a nonnegative number and $U(t, s)$ is a continuous function of (s, t) such that for all $(s, t) \in [a, b] \times [a, b]$,

$$|U(s, t)| \leqq \frac{r}{b - a}$$

where r is a positive number. Then the corresponding mapping $\mathscr{K}$ is a compact mapping which takes the sphere

$$S = [g \in C \,/\, \|g\| \leqq r]$$

into itself.

These examples are the simplest cases of applications of fixed point theorems to nonlinear integral equations. More complicated cases have been dealt with by several writers. For example, W. Pogorzelski [**1**] has used the Schauder Fixed Point Theorem to establish existence theorems for a broad class of singular integral equations. Also the theory of Schmidt operators (§ 8 of Chapter III) may be applied to the study of integral equations. This local theory is an abstraction of a study of integral equations made by Schmidt [**1**] but the abstract local theory can then be applied to a wider class of integral equations than that studied by Schmidt. See Bartle [**1**] and Cronin [**1**; **2**].

In the book of Krasnos'elskiĭ [**1**] several topological methods besides those we deal with here are studied and applied to a variety of concrete problems in integral equations: eigenvalues, branch points, nonlinear spectral analysis as well as existence questions. The Schauder Fixed Point Theorem has also been applied to integral equations in Orlicz spaces by Krasnos'elskiĭ and Rutickiĭ [**1**].

3. **Problems in partial differential equations.** In the sections which follow, we will discuss certain classical problems for particular types of partial differential equations: the Dirichlet problem for elliptic and parabolic equations and the Cauchy problem and mixed problems for hyperbolic equations.

These classical problems have their origins in physical problems. However a student with a purely mathematical viewpoint may well ask why the mathematician should be bound by a tradition from physics. Why should we not, for example, consider the Dirichlet problem for a hyperbolic equation or the Cauchy problem for an elliptic equation. It is not difficult to show that these problems lead to serious difficulties. One important difficulty is that the solution does not depend continuously on the given boundary conditions. For a discussion of these difficulties, see Hadamard [**1**, Book I] and Courant and Hilbert [**2**, pp. 171–179].

The problems we will discuss form a very limited class of problems among the possible problems in partial differential equations. Only certain problems in second order equations will be considered.

Elliptic Differential Equations

First we will describe in some detail the major result of Leray and Schauder [**1**] which is one of the most striking applications of topological theory in function theory. Then the improved version of the Leray-Schauder result given by Nirenberg [**1**] will be described. Finally a number of other applications to elliptic equations will be briefly mentioned.

4. **Statement of the Leray-Schauder-Nirenberg result.** Very roughly, the result says that the Dirichlet problem for the elliptic equation

$$A(x, y, z, z_x, z_y)z_{xx} + B(x, y, z, z_x, z_y)z_{xy} + C(x, y, z, z_x, z_y)z_{yy} = 0 \tag{4.1}$$

where functions A, B, and C are well behaved, has a solution and this solution is well behaved. (The meaning of the terms "Dirichlet problem" and "well behaved" will be described in detail later.) The fact that A, B, and C are functions of z is what makes this problem difficult. For if A, B, and C are independent of z, then it is easy to show that the Dirichlet problem for (4.1) has a unique solution (see Courant and Hilbert [**2**, pp. 276–277]). And the existence of a solution was long ago established by S. Bernstein [**2**]. Whether

(4.1) has a unique solution is still an open question. Here we study only the problem of whether (4.1) has at least one solution if A, B, and C are functions of z. (This is the weaker type of existence question that we may be able to solve by using topological techniques.) The method devised by Leray and Schauder uses the topological degree. However the Schaefer Theorem (Theorem (4.4), Chapter III) is actually sufficient, i.e., the full force of the degree theory is not used. Nirenberg's method requires only the Schauder Fixed Point Theorem.

5. **The Banach spaces in which the Leray-Schauder-Nirenberg result is formulated.** First we introduce certain Banach spaces in which the problems will be staged. These Banach spaces are sets of functions all with domain a subset of the xy-plane. Corresponding definitions may be given for functions with domain a subset of Euclidean n-space where $n > 2$. (See Graves [**1**].) But we will not use such spaces here. Let $\mathscr{D}$ be a bounded connected open set in the xy-plane and let $\mathscr{D}'$ be its boundary. Let m be a non-negative integer and let f be a real-valued function with domain $\mathscr{D}$.

Definition. If $m > 0$, the function f is *of class* C_m on $\mathscr{D}$ (we write: $f \in C_m(\mathscr{D})$) if all the directional derivatives of f up to order m are defined and continuous on $\mathscr{D}$. If $m = 0$, function f is *of class* C_m *on* $\mathscr{D}$ if f is continuous on $\mathscr{D}$.

Definition. Let $\alpha \in (0, 1)$. Function f is *α-Hölder continuous on* $\mathscr{D}$ if

$$M_\alpha(f) = \operatorname*{lub}_{(p,q) \in \mathscr{D} \times \mathscr{D}} \frac{|f(p) - f(q)|}{|p - q|^\alpha}$$

is finite. The number $M_\alpha(f)$ is called the *α-Hölder constant* of f. (In this definition, we include the case in which $\mathscr{D}$ is unbounded.)

For each $f \in C_0(\mathscr{D})$, let

$$M_0(f) = \operatorname*{lub}_{p \in \mathscr{D}} |f(p)|$$

and

$$M_{0+\alpha}(f) = M_\alpha(f).$$

Let $D^m f$ denote any directional derivative of mth order of function f. For each $f \in C_m(\mathscr{D})$ where $m > 0$, define:

$$M_m(f) = \operatorname{lub} M_0(D^m f),$$
$$M_{m+\alpha}(f) = \operatorname{lub} M_\alpha(D^m f),$$

where the lubs are taken over all directional derivatives of order m. Finally for $f \in C_m(\mathscr{D})$ where $m \geqq 0$, define:

$$\|f\|_m = \sum_{j=0}^{m} M_j(f)$$

and

$$\|f\|_{m+\alpha} = \sum_{j=0}^{m} M_j(f) + M_{m+\alpha}(f);$$

and let

$$C_m(\overline{\mathscr{D}}) = [f \in C_m(\mathscr{D}) \mid \|f\|_m < \infty]$$

and

$$C_{m+\alpha}(\overline{\mathscr{D}}) = [f \in C_m(\mathscr{D}) \mid \|f\|_{m+\alpha} < \infty].$$

For simplicity, denote $C_{0+\alpha}(\overline{\mathscr{D}})$ by $C_\alpha(\overline{\mathscr{D}})$.

(5.1) LEMMA. *If $f \in C_{m+\alpha}(\overline{\mathscr{D}})$, then each $D^m f$ has a continuous extension F on $\overline{\mathscr{D}}$, i.e., there is a real-valued function F with domain $\overline{\mathscr{D}}$ such that F is continuous on $\overline{\mathscr{D}}$ and if $p \in \mathscr{D}$, then $F(p) = D^m f(p)$.*

PROOF. Since $f \in C_{m+\alpha}(\overline{\mathscr{D}})$, then $D^m f$ is uniformly continuous on $\mathscr{D}$ and hence by a standard argument using the Cauchy condition, the existence of extension F follows. ∎

(5.2) LEMMA. *The set $C_{m+\alpha}(\overline{\mathscr{D}})$ is a Banach space under the norm $\|f\|_{m+\alpha}$.*

PROOF. It is easy to see that $C_{m+\alpha}(\overline{\mathscr{D}})$ is a linear space and that $\|f\|_{m+\alpha}$ has the properties of a norm. It remains only to prove that $C_{m+\alpha}(\overline{\mathscr{D}})$ is complete under this norm. Suppose that $\{f_n\}$ is a sequence in $C_{m+\alpha}(\overline{\mathscr{D}})$ such that

$$\lim_{k,\, n \to \infty} \|f_k - f_n\|_{m+\alpha} = 0. \tag{5.3}$$

Since $f_n \in C_{m+\alpha}(\overline{\mathscr{D}})$ then the directional derivatives of f_n up to order m are all continuous and converge uniformly on $\mathscr{D}$. Hence each converges to a continuous function. Suppose f is the function to which the functions f_n converge. Since the sequences $D^j f_n$ $(j = 1, \cdots, m)$ all converge uniformly, then $D^j f_n$ converges to $D^j f$. Now we must show that $M_{m+\alpha}(f)$ is finite and finally, that

$$\lim_{n \to \infty} M_{m+\alpha}(f_n - f) = 0. \tag{5.4}$$

First

$$\begin{aligned} \frac{|D^m f(p) - D^m f(q)|}{|p - q|^\alpha} &\leq \frac{|D^m f(p) - D^m f_n(p)| + |D^m f_n(p) - D^m f_n(q)|}{|p - q|^\alpha} \\ &\quad + \frac{|D^m f_n(q) - D^m f(q)|}{|p - q|^\alpha} \\ &\leq M_m(f_n) + 2\,\frac{\varepsilon |p - q|^\alpha}{|p - q|^\alpha} \end{aligned} \tag{5.5}$$

if n is sufficiently large. From (5.3) it follows by the usual arguments that $\{\|f_n\|_{m+\alpha}\}$ is bounded and hence $\{M_{m+\alpha}(f_n)\}$ is a bounded set. Hence $M_{m+\alpha}(f)$ is finite.

Now we prove (5.4). By hypothesis, for arbitrary p and q, and for all sufficiently large k and n,

$$|D^m(f_k - f_n)(p) - D^m(f_k - f_n)(q)| < \varepsilon|p - q|^\alpha; \tag{5.6}$$

and for all sufficiently large n and for arbitrary p,

$$|D^m(f_n - f)(p)| < \varepsilon. \tag{5.7}$$

We need to prove: for arbitrary p, q and for sufficiently large n,

$$|D^m(f_n - f)(p) - D^m(f_n - f)(q)| < \varepsilon|p - q|^\alpha. \tag{5.8}$$

Suppose (5.8) is not true. Then there is a positive constant ε' and a sequence $\{p_r, q_r\}$ of pairs of points in D and a subsequence $\{f_{n_r}\}$ of $\{f_n\}$ such that

$$|D^m(f_{n_r} - f)(p_r) - D^m(f_{n_r} - f)(q_r)| \geqq \varepsilon'|p_r - q_r|^\alpha. \tag{5.9}$$

For fixed n_{r_0} sufficiently large and for all $n_r > n_{r_0}$, we have by (5.6):

$$|D^m(f_{n_r} - f_{n_{r_0}})(p_{r_0}) - D^m(f_{n_r} - f_{n_{r_0}})(q_{r_0})| \leqq \frac{\varepsilon'}{2}\,|p_{r_0} - q_{r_0}|^\alpha. \tag{5.10}$$

From (5.9) with $n_r = n_{r_0}$ and (5.10), it follows that for all sufficiently large n_r,

$$|D^m(f_{n_r} - f)(p_{r_0}) - D^m(f_{n_r} - f)(q_{r_0})| \geqq \frac{\varepsilon'}{2}\,|p_{r_0} - q_{r_0}|^\alpha.$$

Since p_{r_0}, q_{r_0} are fixed, this contradicts (5.7). ∎

Next we prove a lemma which is essential for our purposes since it will yield the compactness property of the transformations we work with.

(5.11) **Lemma.** *If $0 < \gamma < \alpha < 1$, the unit sphere S of $C_{m+\alpha}(\overline{\mathscr{D}})$ is compact as a subset of $C_{m+\gamma}(\overline{\mathscr{D}})$, i.e., each bounded infinite subset of $C_{m+\alpha}(\overline{\mathscr{D}})$ has a limit point (in the topology of $C_{m+\gamma}(\overline{\mathscr{D}})$) in $C_{m+\gamma}(\overline{\mathscr{D}})$ and this limit point is an element of $C_{m+\alpha}(\mathscr{D})$.*

Proof. Suppose the sequence $\{f_n\}$ is contained in $C_{m+\alpha}(\overline{\mathscr{D}})$ and

$$\|f_n\|_{m+\alpha} \leqq 1$$

for all n. By Ascoli's Theorem (Theorem (2.5)), there is a subsequence $\{f_{n_k}\}$ of $\{f_n\}$ which converges uniformly to a function f and such that $D^j f_{n_k}$ converges uniformly to $D^j f$ for $j = 1, \cdots, m$. This follows because if $\|f_n\|_{m+\alpha} \leqq 1$, then for all n,

$$M_j[f_n] \leqq 1$$

and

$$M_{m+\alpha}(f_n) \leqq 1.$$

Hence $\{D^j f_n\}$ is equicontinuous and bounded for $j = 1, \cdots, m$.

Now we must show: $f \in C_{m+\alpha}(\mathscr{D})$ and

$$\lim_{k\to\infty} M_{m+\gamma}(f_{n_k} - f) = 0. \tag{5.12}$$

To prove that $f \in C_{m+\alpha}(\overline{\mathscr{D}})$, the same kind of argument as embodied in inequality (5.5) can be used. To prove (5.12) we use the inequality:

$$[M_\gamma(g)]^\alpha \leqq [2M_0(g)]^{\alpha-\gamma}[M_\alpha(g)]^\gamma. \tag{5.13}$$

Assuming (5.13) is true, we then prove (5.12) by taking

$$g = f_{n_k} - f.$$

The numbers $M_\alpha(f_{n_k} - f)$ form a bounded set and

$$\lim_{k\to\infty} M_0(f_{n_k} - f) = 0.$$

Finally, we prove (5.13):

$$\begin{aligned}
[M_\gamma(g)]^\alpha &= \left\{\text{lub}\,\frac{|g(p) - g(q)|}{|p - q|^\gamma}\right\}^\alpha \\
&= \text{lub}\,\frac{|g(p) - g(q)|^\alpha}{|p - q|^{\gamma\alpha}} \\
&\leqq \left\{\text{lub}\,\frac{|g(p) - g(q)|^\gamma}{|p - q|^{\gamma\alpha}}\right\}[2M_0(g)]^{\alpha-\gamma} \\
&= \left\{\text{lub}\,\frac{|g(p) - g(q)|}{|p - q|^\alpha}\right\}^\gamma [2M_0(g)]^{\alpha-\gamma} \\
&= \{M_\alpha g\}^\gamma[2M_0(g)]^{\alpha-\gamma}. \quad \blacksquare
\end{aligned}$$

Now we assume that the boundary $\mathscr{D}'$ of $\mathscr{D}$ is the point set (i.e., the set of image points) of a curve Γ of finite length represented by

$$x = x(s), \qquad y = y(s)$$

where s is arclength on Γ. By simplified versions of the definitions and arguments in the preceding section we have: the set of functions $\phi(s)$ defined on $\mathscr{D}'$ as functions of arclength s which have continuous derivatives of mth order $\phi^{(m)}(s)$ such that

$$M_{m+\alpha}\phi(s) = \operatorname*{lub}_{s_1,\, s_2 \in \mathscr{D}'} \frac{|\phi^{(m)}(s_1) - \phi^{(m)}(s_2)|}{|s_1 - s_2|^\alpha} < \infty,$$

where $\alpha \in (0, 1)$, form a Banach space $C_{m+\alpha}(\mathscr{D}')$ under the norm

$$\|\phi\| = \max_{\mathscr{D}'} |\phi(s)| + \max_{\mathscr{D}'} |\phi^{(1)}(s)| + \cdots + \max_{\mathscr{D}'} |\phi^{(m)}(s)| + M_{m+\alpha}(\phi).$$

Also we define:

$$C_m(\mathscr{D}') = [\phi(s) \,/\, \max_{\mathscr{D}'} |\phi(s)| + \sum_{j=1}^{m} \max_{\mathscr{D}'} |\phi^{(j)}(s)| < \infty].$$

6. **The Schauder Existence Theorem.** Application of the topological techniques of Chapter III requires two steps: first the problem must be formulated in terms of a compact mapping and secondly we must either show that the mapping is an into mapping on some convex bounded set (if the Schauder Fixed Point Theorem is to be applied) or obtain a priori estimates (if the Schaefer Theorem (Theorem (4.4), Chapter III) or the Leray-Schauder degree theory is to be applied). To formulate the present problem in terms of a compact mapping we will use an existence theorem for linear elliptic equations due to Schauder. In order to state this theorem, we need first to impose a stronger hypothesis on $\mathscr{D}'$.

(6.1) DEFINITION. Boundary *$\mathscr{D}'$ is in $C'_{j+\mu}$* where j is a positive integer and $\mu \in (0, 1)$ if the following conditions hold:

(1) at each point $P_0 \in \mathscr{D}'$, there is a normal;
(2) if the origin is taken at P_0 with the x-axis along the normal, then there is a fixed positive number σ independent of P_0 such that the points of $\mathscr{D}'$ with $|x| < \sigma$ and $|y| < \sigma$ belong to the graph of a function

$$x = g(y)$$

such that g has a jth derivative which is μ-Hölder continuous on the set

$$[y \,/\, |y| < \sigma].$$

(6.2) SCHAUDER EXISTENCE THEOREM. *Let the boundary $\mathscr{D}'$ of $\mathscr{D}$ be in $C'_{2+\mu}$. Let*

$$a(x, y)z_{xx} + b(x, y)z_{xy} + c(x, y)z_{yy} = \rho(x, y) \tag{6.3}$$

be a linear equation on $\mathscr{D}$ satisfying the following conditions:
(1) *$a, b, c \in C_\mu(\overline{\mathscr{D}})$ where $0 < \mu < 1$;*
(2) *$\rho \in C_\mu(\overline{\mathscr{D}})$;*
(3) *there is a positive constant m such that for all real ξ, η and all $(x, y) \in \mathscr{D}$:*

$$a(x, y)\xi^2 + b(x, y)\xi\eta + c(x, y)\eta^2 \geqq m(\xi^2 + \eta^2),$$

i.e., equation (6.3) *is elliptic.*
Let $\phi(s) \in C_{2+\mu}(\mathscr{D}')$.
Conclusion: there is a unique solution $z(x, y)$ of (6.3) *such that $z(x, y) = \phi(s)$ on $\mathscr{D}'$ (we will say $z(x, y)$ has boundary value $\phi(s)$ on $\mathscr{D}'$) and $z(x, y) \in C_{2+\mu}(\overline{\mathscr{D}})$. Also there exists a positive number k which depends only on $\|a\|_\mu$, $\|b\|_\mu$, $\|c\|_\mu$, m and $\mathscr{D}$ such that*

$$\|z\|_{2+\mu} \leqq k\{\|\phi\|_{2+\mu} + \|\rho\|_\mu\}.$$

We have stated the Schauder Existence Theorem only for the two-dimensional case. It holds also for the n-dimensional case and in the more general case in which terms of the form $d(x, y)z_x$ and $e(x, y)z_y$ and $f(x, y)z$

(with $f(x, y) \leqq 0$ for all (x, y)) appear in equation (6.3). For simplicity of notation, we deal only with the 2-dimensional case, and we do not use the more general equation containing terms $d(x, y)z_x + e(x, y)z_y + f(x, y)z$.

The Schauder Existence Theorem is at the basis of both Leray-Schauder and Nirenberg proofs. The proof is lengthy and complicated and Schauder's original proof (see Schauder [**3**]) contains gaps. A thorough reworking of the proof has been given by Graves [**1**]. Also the "interior estimates" have been obtained for more general elliptic systems by Douglis and Nirenberg [**1**]. For further references and discussion of this subject, see Miranda [**1**] and Barrar [**2**].

7. The Leray-Schauder method. First we state precisely the problem studied by Leray and Schauder. We consider the equation

$$A(x, y, z, z_x, z_y)z_{xx} + B(x, y, z, z_x, z_y)z_{xy} + C(x, y, z, z_x, z_y)z_{yy} = 0 \tag{7.1}$$

where functions $A(x, y, z, p, q)$, $B(x, y, z, p, q)$ and $C(x, y, z, p, q)$ are defined for all $(x, y) \in \mathscr{D}$, a domain with boundary $\mathscr{D}'$ in $C'_{2+\alpha}(0 < \alpha < 1)$, and all real z, p, q. Functions A, B, and C have α-Hölder continuous second derivatives in all variables for $(x, y) \in \mathscr{D}$ and all real z, p, q; and there is a positive number m such that for all real ξ and η, all $(x, y) \in \mathscr{D}$ and all z, p, q,

$$A\xi^2 + B\xi\eta + C\eta^2 \geqq m(\xi^2 + \eta^2).$$

(That is, equation (7.1) is elliptic on $\mathscr{D}$.) Now let

$$\phi(s) \in c_{3+\alpha}(\mathscr{D}').$$

The Dirichlet problem we study is this: is there a solution $z(x, y)$ of (7.1) in D which has boundary value $\phi(s)$ on $\mathscr{D}'$? (The prototype of this problem is the classical problem of potential theory.) We will show that there is at least one such solution $z(x, y)$ and $z(x, y) \in C_{2+\alpha}(\overline{\mathscr{D}})$.

The technique devised by Leray and Schauder consists in studying first the equation:

$$A(x, y, z, z_x, z_y)U_{xx} + B(x, y, z, z_x, z_y)U_{xy} + C(x, y, z, z_x, z_y)U_{yy} = 0 \tag{7.2}$$

where $z(x, y)$ is a given element of $C_{2+\beta}(\overline{\mathscr{D}})$ where $0 < \beta < \alpha < 1$. Then we may apply the Schauder Existence Theorem (6.2) and conclude that there is a solution $U(z)$ of equation (7.2) such that $U(z)$ has boundary value $\phi(s)$ on $\mathscr{D}'$ and $U(z) \in C_{2+\alpha}(\overline{\mathscr{D}})$. Since $\beta < \alpha$, then by Lemma (5.11), the transformation

$$U : z \to U(z)$$

takes a bounded set in $C_{2+\beta}(\overline{\mathscr{D}})$ into a compact set in $C_{2+\beta}(\mathscr{D})$. Also transformation U is a continuous mapping from $C_{2+\beta}(\overline{\mathscr{D}})$ into $C_{2+\beta}(\overline{\mathscr{D}})$ as follows easily from the inequality in the conclusion of the Schauder Existence Theorem (6.2). Hence U is a compact mapping. From the definition of $U(z)$, it is clear that to solve our Dirichlet problem, it is sufficient to find a solution in $C_{2+\beta}(\overline{\mathscr{D}})$ of the functional equation:

$$z = U(z) \tag{7.3}$$

or

$$z - U(z) = \bar{0}.$$

Since U is compact, the LS degree of $I - U$ is defined and if it can be shown that

$$d[I - U, S, \bar{0}] \neq 0 \tag{7.4}$$

where I is the identity mapping and S is a sphere in $C_{2+\beta}(\overline{\mathscr{D}})$, then by Property 1 of the LS degree in § 5 of Chapter III, the problem is solved. In order to establish (7.4), Leray and Schauder study the homotopy (see § 5 of Chapter III)

$$z - U(z, t)$$

where $U(z, t)$ is the solution of (7.2) such that on $\mathscr{D}'$,

$$U(z, t) = t\phi(s),$$

where $0 \leqq t \leqq 1$. (Since the number k in the conclusion of the Schauder Theorem depends only on $\|a\|_\mu$, $\|b\|_\mu$, $\|c\|_\mu$, m and $\mathscr{D}$, it is clear that $U(z, t)$ is a homotopy.) Now if $t = 0$, then

$$U(z, 0) = \bar{0}$$

for all z. Hence if $t = 0$ and S is any sphere in $C_{2+\beta}(\overline{\mathscr{D}})$ then

$$d[I - U(z, 0), S, \bar{0}] = d[I, S, \bar{0}].$$

But

$$d[I, S, \bar{0}] = +1.$$

Thus if we can find a sphere S in $C_{2+\beta}(\overline{\mathscr{D}})$ such that for all $t \in [0, 1]$, the equation

$$z - U(z, t) = 0$$

has no solutions except possibly in the interior of S, then by Theorem (7.3) of Chapter III, it follows that

$$d[I - U(z, t), S, \bar{0}] = +1.$$

Hence (7.3) has a solution in S or the Dirichlet problem for equation (7.1) has a solution which is in $C_{2+\beta}(\overline{\mathscr{D}})$ and also in $C_{2+\alpha}(\overline{\mathscr{D}})$ with boundary values $\phi(s)$ where $\phi(s)$ is the given function in $c_{3+\alpha}(\mathscr{D}')$. Note that we can draw no conclusions about the number of distinct solutions. Note also that we could use Schaefer's Theorem (Theorem (4.4) in Chapter III) instead of the Leray-Schauder theory, i.e., the full force of the degree theory is not used.

Showing the existence of the sphere S described in the preceding paragraph is by far the most difficult part of the application of the Leray-Schauder technique to analysis. We will obtain a priori bounds (cf. Definition (7.1) of Chapter III) for the original equation (7.1), that is, we will obtain a statement of the form: there is a positive constant K such that if equation (7.1) has a solution in $C_{2+\beta}(\overline{\mathscr{D}})$ for any of the boundary values $t\phi$ where $t \in [0, 1]$ then the norm of each of these solutions (regarded as elements of $C_{2+\beta}(\overline{\mathscr{D}})$) is less than K.

We obtain the a priori bounds from a sequence of theorems which will be stated without proof.

(7.5) THEOREM. (The Maximum Principle). *Let*

$$a(x, y)z_{xx} + b(x, y)z_{xy} + c(x, y)z_{yy} = 0 \tag{7.6}$$

be a linear equation on $\mathscr{D}$ *satisfying the following conditions*:

(1) $a(x, y), b(x, y), c(x, y)$ *are continuous functions of* x *and* y *on* $\mathscr{D} \cup \mathscr{D}'$;
(2) *for all* $(x, y) \in \mathscr{D} \cup \mathscr{D}'$,

$$b^2 - 4ac < 0.$$

If $z(x, y)$ *is a solution of* (7.6) *such that* $z(x, y)$ *has continuous second derivatives in* $\mathscr{D}$ *and is continuous on* $\mathscr{D} \cup \mathscr{D}'$ *and if the boundary value of* $z(x, y)$ *on* $\mathscr{D}'$ *is* $\phi(s)$, *then*

$$\max_{(x, y) \in \mathscr{D} \cup \mathscr{D}'} |z(x, y)| \leqq \max_{s \in \mathscr{D}'} |\phi(s)|.$$

PROOF. See E. Hopf [**1**] or Miranda [**1**].

DEFINITION. Let $\phi(s)$ be a real-valued continuous function with domain $\mathscr{D}'$ and let $\mathscr{D}'$ be the point set of a curve of finite length which is described by

$$x = x(s), \qquad y = y(s).$$

The curve C described by:

$$x = x(s), \qquad y = y(s), \qquad z = \phi(s)$$

satisfies a triangle condition with constant Δ, where Δ is a positive number, if for each plane

$$z = \alpha x + \beta y + \gamma$$

that passes through three distinct points of C, the following inequality holds:

$$\sqrt{\alpha^2 + \beta^2} \leqq \Delta.$$

(7.7) THEOREM. *Suppose that $x(s)$, $y(s)$ have continuous second derivatives and suppose that the curvature of the curve described by*

$$x = x(s), \qquad y = y(s)$$

(of which $\mathscr{D}'$ is the corresponding point set) is everywhere positive. Assume further that function $\phi(s)$ with domain $\mathscr{D}'$ has a continuous second derivative and that there is a positive constant L such that for all s,

$$\max_{\mathscr{D}'} \{|\phi(s)| + |\phi'(s)| + |\phi''(s)|\} \leqq L.$$

Then the curve C represented by

$$x = x(s), \qquad y = y(s), \qquad z = \phi(s)$$

satisfies a triangle condition with constant Δ where Δ depends only on L for a given curve described by $x = x(s)$, $y = y(s)$.

PROOF. See Schauder [**2**].

DEFINITION. Let $z(x, y)$ be a continuous function on $\mathscr{D} \cup \mathscr{D}'$. If for any domain $\mathfrak{A}$ such that $\mathfrak{A} \subset \mathscr{D}$ and such that on the boundary of $\mathfrak{A}$,

$$z(x, y) = \alpha x + \beta y + \gamma$$

where α, β, γ are real numbers, it follows that

$$z(x, y) = \alpha x + \beta y + \gamma$$

in the interior of $\mathfrak{A}$, then $z(x, y)$ is a *saddle function* on $\mathscr{D}$. (Note that the function $z(x, y)$ is a saddle function if and only if the Gauss curvature of the surface described by $z(x, y)$ is nonpositive.)

(7.8) THEOREM. *If $z(x, y)$ satisfies equation* (7.6) *in the domain $\mathscr{D}$ described in Theorem* 7.5, *then $z(x, y)$ is a saddle function on $\mathscr{D}$.*

PROOF. This follows from a simple computation. See Nirenberg [**1**, p. 140] or Schauder [**2**].

(7.9) THEOREM (RADO'S THEOREM). *Let $\mathscr{D}$ be a convex bounded domain, $\mathscr{D}'$ the point set of a curve represented by $x = x(s)$, $y = y(s)$ and suppose $z(x, y)$ is a saddle function on $\mathscr{D}$ such that $z(x, y)$ is continuous on $\mathscr{D} \cup \mathscr{D}'$ and the curve*

$$x = x(s), \qquad y = y(s), \qquad z = z/\mathscr{D}'$$

satisfies a triangle condition with constant Δ. Then $z(x, y)$ satisfies in $\mathscr{D}$ the Lipschitz condition:

$$|z(x_1, y_1) - z(x_2, y_2)| \leqq \Delta \sqrt{(x_1 - x_2)^2 + (y_1 - y_2)^2}.$$

PROOF. This theorem was proved by Rado [**1**].

A simple and elegant proof was given by von Neumann [**1**]. (Von Neumann's hypothesis is weaker than the condition that $z(x, y)$ be a saddle function.)

(7.10) THEOREM. *Let $\mathscr{D}$ be a convex bounded domain such that $\mathscr{D}'$ is the point set associated with a curve represented by*

$$x = x(s), \qquad y = y(s)$$

where $x(s)$, $y(s)$ have continuous second derivatives. In the equation

$$A(x, y, z, p, q)z_{xx} + B(x, y, z, p, q)z_{xy} + C(x, y, z, p, q)z_{yy} = 0, \tag{7.11}$$

assume:

(1) *functions A, B, C are defined for all z, p, q and for $(x, y) \in \mathscr{D}$;*

(2) *functions A, B, C have continuous second derivatives in all variables for $(x, y) \in \mathscr{D}$ and all z, p, q; and these second derivatives are α-Hölder continuous.*

Let $z(x, y)$ be a solution of (7.11) *which has second derivatives at each point of $\mathscr{D}$ and suppose these second derivatives are α-Hölder continuous in $\mathscr{D}$. Suppose also that $z(x, y)$ has boundary value $\phi(s)$ on $\mathscr{D}'$ where $\phi(s) \in c_{3+\alpha}(\mathscr{D}')$. If there are positive constants μ_1 and μ_2 such that for all $(x, y) \in \mathscr{D}$,*

$$|z(x, y)| + |D^1 z| < \mu_1$$

and

$$\|\phi\|_{3+\alpha} \leqq \mu_2,$$

then there exist constants C_1 and C_2 which depend only on μ_1 and μ_2 such that

$$|D^2 z| \leqq C_1$$

for all $(x, y) \in \mathscr{D}$ and

$$\|D^2 z\|_\alpha \leqq C_2.$$

PROOF. The proof of this theorem is given by Schauder [**2**] although not in full detail because it is an extension of a proof given by S. Bernstein. Schauder remarks ([**2**] footnote (14)) that it would probably be sufficient to assume that $\phi(s)$ has α-Hölder continuous second derivatives.

Now we combine these results to get the a priori bounds.

(7.12) THEOREM. *Let $\mathscr{D}$ be a bounded convex domain in the xy-plane such that $\mathscr{D}'$ is the point set of a curve Γ which is represented by:*

$$x = x(s), \qquad y = y(s)$$

where $x(s)$ and $y(s)$ have continuous second derivatives and such that Γ has positive curvature everywhere. Let $\phi(s) \in c_{3+\alpha}(\mathscr{D}')$ be given. Then if $z(x, y)$ is a solution of (7.1) *which has boundary value $k\phi(s)$ on $\mathscr{D}'$ where $k \in (0, 1)$ and which has α-Hölder continuous second derivatives in $\mathscr{D}$, there is a positive constant K which is independent of k and which is such that*

$$\|z\|_{2+\alpha} \leqq K.$$

PROOF. Substituting $z(x, y)$ into A, B, C in equation (7.11) we get functions $a(x, y)$, $b(x, y)$, $c(x, y)$ and equation (7.1) becomes equation (7.6). Applying Theorem (7.5) (the Maximum Principle), we obtain:

$$(7.13)\qquad \max_{(x,y)\in\mathscr{D}\cup\mathscr{D}'} |z(x, y)| \leqq \max_{s\in\mathscr{D}'} |k\phi(s)| \leqq \|\phi(s)\|_{3+\alpha}.$$

By Theorem (7.7), the curve in 3-space represented by

$$x = x(s), \qquad y = y(s), \qquad z = k\phi(s)$$

satisfies a triangle condition with constant Δ where Δ depends only on $\|\phi\|_{3+\alpha}$. Since $z(x, y)$ satisfies equation (7.6) then by Theorem (7.8), function $z(x, y)$ is a saddle function. Hence we may apply Theorem (7.9) (Rado's Theorem) and conclude that the first derivatives of z satisfy the condition

$$(7.14)\qquad \max_{(x,y)\in\mathscr{D}\cup\mathscr{D}'} |D^1 z| \leqq \Delta.$$

From (7.13) and (7.14), we have by applying Theorem (7.10):

$$\|z\|_{2+\alpha} \leqq K$$

where K depends only on $\|\phi\|_{3+\alpha}$. ∎

We may summarize the results of this section as:

(7.15) THEOREM (LERAY-SCHAUDER THEOREM). *Let $\mathscr{D}$ be a bounded convex domain in the xy-plane such that $\mathscr{D}'$ is the point set of a curve Γ which is represented by*

$$x = x(s), \qquad y = y(s),$$

where $x(s)$ and $y(s)$ have continuous second derivatives, and such that Γ has positive curvature everywhere and $\mathscr{D}'$ is in $C'_{2+\alpha}$ (where $0 < \alpha < 1$). Let

$$(7.16)\quad A(x, y, z, z_x, z_y)z_{xx} + B(x, y, z, z_x, z_y)z_{xy} + C(x, y, z, z_x, z_y)z_{yy} = 0$$

be an equation such that the functions A, B, and C are defined for all $(x, y) \in \mathscr{D}$ and all real z, p, q, and have α-Hölder continuous second derivatives in all variables for $(x, y) \in \mathscr{D}$ and all real z, p, q. Suppose further there is a positive number m such that for all real ξ and η, all $(x, y) \in \mathscr{D}$ and all z, p, q,

$$A\xi^2 + B\xi\eta + C\eta^2 \geqq m(\xi^2 + \eta^2).$$

Suppose

$$\phi(s) \in c_{3+\alpha}(\mathscr{D}').$$

Then (7.16) *has at least one solution $z(x, y)$ such that $z(x, y)$ has boundary value $\phi(s)$ on $\mathscr{D}'$ and such that*

$$z(x, y) \in C_{2+\alpha}(\overline{\mathscr{D}}).$$

8. **The Nirenberg method.** Nirenberg [**1**] applies the technique devised by Leray and Schauder to derive a functional equation of the form (7.3).

However, instead of applying the Schaefer Theorem, Nirenberg uses the Schauder Fixed Point Theorem. In order to use this theorem, it is necessary to select a bounded convex set of functions which is mapped into itself by the transformation U. Obtaining such a set requires again rather lengthy considerations. The theorems used in the Leray-Schauder method are used again except for Theorem (7.10). Instead of Theorem (7.10), some results concerning linear elliptic equations (obtained by Nirenberg [**1**]) are applied. Besides the simpler topological technique used, Nirenberg's method has the advantage of clear and complete exposition.

First we state precisely the problem studied by Nirenberg. Let

(8.1) $A(x, y, z, z_x, z_y)z_{xx} + B(x, y, z, z_x, z_y)z_{xy} + C(x, y, z, z_x, z_y)z_{yy} = 0$

satisfy the following conditions:

(1) the functions $A(x, y, z, p, q)$, $B(x, y, z, p, q)$, and $C(x, y, z, p, q)$ are defined for all real z, p, q and for all $(x, y) \in \mathscr{D}$, a bounded convex domain in the xy-plane with boundary $\mathscr{D}'$ in $C'_{2+\beta}$ $(0 < \beta < 1)$ such that $\mathscr{D}'$ is the point set associated with a curve Γ with positive curvature everywhere;

(2) for each positive number K and for all x, y, z, p, q satisfying the conditions: $(x, y) \in \mathscr{D}$ and $|z|, |p|, |q| \leqq K$, it is assumed that:

(i) functions A, B, C are β-Hölder continuous in x, y, z, p, q where β and the β-Hölder constants depend on K;

(ii) there are positive constants $m(K)$ and $M(K)$, which depend on K, such that for all real ξ, η and $(x, y) \in \mathscr{D}$ and $|z|, |p|, |q| \leqq K$.

$$[M(K)](\xi^2 + \eta^2) \geqq A\xi^2 + B\xi\eta + C\eta^2 \geqq [m(K)](\xi^2 + \eta^2).$$

Let $\phi(s) \in C_{2+\alpha}(\mathscr{D}')$, where $0 < \alpha < 1$, be given. It will be shown that there is a solution $z(x, y)$ of (8.1) in $\mathscr{D}$ such that the boundary value of $z(x, y)$ is $\phi(s)$ and that there is a number $\gamma \in (0, 1)$ such that $z(x, y) \in C_{2+\gamma}(\overline{\mathscr{D}})$.

Applying Theorem (7.5) (the Maximum Principle), Theorems (7.7), (7.8), and (7.9) (Rado's Theorem) yields (see Nirenberg [1, pp. 140–141]):

(8.2) THEOREM. *Let $z(x, y)$ be a solution of the linear elliptic equation*

$$a(x, y)z_{xx} + b(x, y)z_{xy} + c(x, y)z_{yy} = 0 \tag{8.3}$$

where $a(x, y)$, $b(x, y)$, $c(x, y)$ are continuous in $\mathscr{D} \cup \mathscr{D}'$ and $z(x, y)$ is continuous in $\mathscr{D} \cup \mathscr{D}'$ and has continuous second derivatives in $\mathscr{D}$. Let $\phi(s)$ denote the boundary value of $z(x, y)$ on $\mathscr{D}'$ and suppose $\phi(s)$ has a continuous second derivative at each point of $\mathscr{D}'$. Then there is a positive constant k such that

$$\|z\|_1 \leqq k\|\phi\|_2.$$

Next an estimate due to Nirenberg [**1**] is obtained for a Hölder constant for $D^1 z$.

(8.4) THEOREM. *Let $z(x, y) \in C_1(\overline{\mathscr{D}})$ be such that*

$$\|z\|_1 \leqq K$$

where K is a positive constant. Suppose $U(x, y) \in C_2(\overline{\mathscr{D}})$ and $U(x, y)$ is a solution of the linear equation

(8.5) $A(x, y, z, z_x, z_y)u_{xx} + B(x, y, z, z_x, z_y)u_{xy} + C(x, y, z, z_x, z_y)u_{yy} = 0$

where A, B, and C are the coefficients in equation (8.1) with $z(x, y)$ and its first derivatives substituted in A, B, and C, and suppose the boundary value of z on Γ is $\phi(s)$ where $\phi(s) \in c_2(\mathscr{D}')$. Then there exists positive constants $\bar{K}$ and δ (<1) which depend only on K, m, M and $\|\phi\|_2$ such that

$$\|U\|_{1+\delta} \leqq \bar{K}.$$

Now let $S_{1+\delta}$ denote the set of functions z satisfying the conditions:

$$\|z\|_1 \leqq K$$

and

$$\|z\|_{1+\delta} \leqq \bar{K}.$$

By applying the Schauder Existence Theorem (6.2) Nirenberg [**1**] proves:

(8.6) THEOREM. *Let $z(x, y) \in S_{1+\delta}$ and let $A(x, y, z, z_x, z_y) = a(x, y)$, $B(x, y, z, z_x, z_y) = b(x, y)$, $C(x, y, z, z_x, z_y) = c(x, y)$. Suppose $\phi(s) \in c_{2+\alpha}(\mathscr{D}')$. Then there is a unique solution $U(x, y) \in C_2(\overline{\mathscr{D}})$ of*

$$a(x, y)z_{xx} + b(x, y)z_{xy} + c(x, y)z_{yy} = 0$$

which has the boundary value $\phi(s)$. Also there exist positive constants $\bar{H}$ and γ depending only on K, M, m, $\bar{K}$, δ, H, α and $\|\phi\|_{2+\alpha}$ such that

$$U \in C_{2+\gamma}(\overline{\mathscr{D}})$$

and

$$\|U\|_{2+\gamma} \leqq \bar{H}.$$

Now the set $S_{1+\delta}$ is a bounded convex set in $C_{1+\delta}(\overline{\mathscr{D}})$. Let U be the transformation $U: z \to U(x, y)$. Theorem (8.6) shows that the operator U is defined for $z(x, y) \in S_{1+\delta}$. Theorem (8.4) shows that U maps $S_{1+\delta}$ into itself. From the Schauder Existence Theorem (6.2), it follows that $U(z)$ is a continuous mapping of $S_{1+\delta}$ into itself and the estimate

$$\|U\|_{2+\gamma} \leqq \bar{H}$$

shows that the set $U(S_{1+\delta})$ is compact in $C_{1+\delta}(\overline{\mathscr{D}})$. Hence by the Schauder Fixed Point Theorem the equation

$$z = U(z)$$

has a solution $z(x, y)$ and this solution is in $C_{2+\gamma}(\overline{\mathscr{D}})$ because $U(S_{1+\delta}) \subset C_{2+\gamma}(\overline{\mathscr{D}})$. Thus we obtain:

(8.7) NIRENBERG'S THEOREM. *Let*

(8.1) $A(x, y, z, z_x, z_y)z_{xx} + B(x, y, z, z_x, z_y)z_{xy} + C(x, y, z, z_x, z_y)z_{yy} = 0$

satisfy the hypothesis described at the beginning of § 8 *on a domain* $\mathscr{D}$ *with the properties described at the beginning of* § 8. *Let* $\phi(s) \in c_{2+\alpha}(\mathscr{D}')$. *Then there exists* $\gamma \in (0, 1)$ *and* $z(x, y) \in C_{2+\gamma}(\overline{\mathscr{D}})$ *such that* $z(x, y)$ *is a solution of* (8.1) *with boundary value* $\phi(s)$ *on* $\mathscr{D}'$.

9. **Some other work in elliptic equations.** We describe briefly some other studies of elliptic equations which use degree or fixed point theory. First, Cordes [**1**] has, with the aid of special hypotheses on the coefficients a, b, c, extended the Leray-Schauder-Nirenberg result to the n-dimensional case with $n > 2$. An extension to the n-dimensional case can be obtained from work of Nirenberg [**2**] (see Nirenberg's review of Cordes' paper in the *Mathematical Reviews*, Vol. 19, pp. 961–962), but Cordes uses weaker hypotheses.

Many problems in elliptic differential equations can be translated into integral equations which can then be studied by using the Schauder Fixed Point Theorem or the Leray-Schauder theory. Such a translation can be carried out by using Green's function. Leray and Schauder [**1**, Part IV] pointed out that if $G(x, y; \xi, \eta)$ is the harmonic Green's function corresponding to a domain $\mathscr{D}$ in the xy-plane, then finding those solutions in $\mathscr{D}$ of the equation

$$z_{xx} + z_{yy} = f(x, y, z, z_x, z_y)$$

(where f is continuous in all its variables) which vanish on the boundary of $\mathscr{D}$ is equivalent to finding the functions $\zeta(x, y)$ which satisfy the equation

$$\zeta(x, y) = f\left\{x, y, \iint G(x, y; \xi, \eta)\zeta(\xi, \eta)\, d\xi\, d\eta; \frac{\partial}{\partial x}\iint G\zeta\, d\xi\, d\eta; \frac{\partial}{\partial y}\iint G\zeta\, d\xi\, d\eta\right\}. \tag{9.1}$$

This result is obtained by writing

$$z_{xx} + z_{yy} = \zeta(x, y)$$

and using the standard properties of the Green's function. By extensions of the arguments used in § 1 on integral equations, we can show that the transformation defined by the expression on the right in (9.1) is completely continuous in the Banach space of functions $\zeta(x, y)$ which are bounded and measurable on $\mathscr{D}$ and with the norm

$$\|\zeta(x, y)\| = \operatorname*{lub}_{(x, y) \in \mathscr{D}} |\zeta(x, y)|.$$

An application of this approach is made by Duff [**1**] who obtains a priori bounds for a particular class of quasilinear elliptic equations and applies the Schaefer Theorem (Theorem (4.4) of Chapter III).

An application of particular interest was made by Heinz [**1**] who used Schaefer's Theorem to treat the difficult problem of the existence of surfaces of constant mean curvature. Since this problem does not, in general, have a unique solution, a topological approach which does not involve considerations of uniqueness is a natural one to use. Finn and Gilbarg [**1**] have used Schaefer's Theorem to establish the existence of certain subsonic flows in 3-space, i.e., to prove the existence of solutions of certain nonlinear elliptic equations outside a "nice" closed surface in 3-space.

Duff, Heinz, and Finn and Gilbarg describe their work in terms of the Leray-Schauder theory. Specifically the theory used is a simple application of Theorem (7.3) of Chapter III. However, this application is exactly Schaefer's Theorem. Thus the concept of the Leray-Schauder degree is actually not needed in these applications.

Another kind of application has been made by Mann and Blackburn [**1**] who study the Laplace equation with a nonlinear boundary condition. By use of a Fourier transform the problem is translated into the study of a nonlinear integral equation which is solved by using the Schauder Fixed Point Theorem. Also Nitsche [**1**] has used the Schauder Fixed Point Theorem to solve linear boundary value problems for quasilinear systems.

10. **Local study of elliptic differential equations.** The Schauder Existence Theorem (Theorem (6.2)) also makes it possible to apply the theory of the Schmidt operator developed in § 8 of Chapter III to an in-the-small study of elliptic equations. For simplicity of notation, we will describe the technique for the case of two independent variables. The same technique is valid in the case of n independent variables although more complicated. (As previously pointed out, the Schauder Existence Theorem holds for $n > 2$.) The technique was introduced by Schauder and applied to this problem by Cronin [**2**; **3**].

Let $\mathscr{D}$ be a bounded connected open set in the xy-plane the boundary $\mathscr{D}'$ of which is in $C'_{2+\alpha}$ where $0 < \alpha < 1$. Consider the equation

$$F(x, y, z, p, q, r, s, t) = \psi(x, y) \tag{10.1}$$

where p, q, r, s, t denote $z_x, z_y, z_{xx}, z_{xy}, z_{yy}$ and F has α-Hölder continuous third derivatives for $(x, y) \in \mathscr{D}$ and p, q, r, s, t in any bounded region and

$$\psi(x, y) \in C_\alpha(\overline{\mathscr{D}}).$$

Now assume that (10.1) has an initial solution in $\overline{\mathscr{D}}$ which is elliptic relative to F. That is, we assume: there exist $\psi_0(x, y) \in C_\alpha(\overline{\mathscr{D}})$ and $\phi_0(s) \in c_{2+\alpha}(\mathscr{D}')$ such that if

$$\psi(x, y) = \psi_0(x, y),$$

then equation (10.1) has a solution $z_0(x, y) \in C_{2+\alpha}(\overline{\mathscr{D}})$ whose boundary value is $\phi_0(s)$; and solution $z_0(x, y)$ is elliptic relative to F, i.e.,

$$\left(\frac{\partial F}{\partial s}\right)^2 - 4\left(\frac{\partial F}{\partial r}\right)\left(\frac{\partial F}{\partial t}\right) < 0$$

if z_0 and its derivatives are substituted for z and its derivatives in $\partial F/\partial s$, $\partial F/\partial r$, $\partial F/\partial t$.

The problem we study is this: if ε is sufficiently small, do there exist solutions $z(x, y) \in C_{2+\alpha}(\overline{\mathscr{D}})$ of equation (10.1) for given $\psi \in C_\alpha(\overline{\mathscr{D}})$ and given boundary value

$$\phi(s) \in c_{2+\alpha}(\mathscr{D}')$$

such that

$$\|\psi - \psi_0\|_\alpha < \varepsilon$$

and

$$\|\phi - \phi_0\|_{2+\alpha} < \varepsilon$$

and how many such solutions are there? We will show in outline how this problem can be phrased in such a form that the theory of Schmidt operators described in § 5 of Chapter III is applicable.

For simplicity, we assume that functions ψ_0 and ϕ_0 are both identically zero. Then (10.1) can be written in the form:

$$\begin{aligned}
\psi(x, y) = {} & F(x, y, z, p, q, r_0, s_0, t_0) \\
& + (r - r_0)\frac{\partial F}{\partial r}(x, y, z_0, p_0, q_0, r_0, s_0, t_0) \\
& + (s - s_0)\frac{\partial F}{\partial s}(x, y, z_0, p_0, q_0, r_0, s_0, t_0) \\
& + (t - t_0)\frac{\partial F}{\partial t}(x, y, z_0, p_0, q_0, r_0, s_0, t_0) \\
& + (r - r_0)\int_0^1 \biggl\{\frac{\partial F}{\partial r}[x, y, z, p, q, r_0 + \lambda(r - r_0), \\
& \qquad s_0 + \lambda(s - s_0), t_0 + \lambda(t - t_0)] \\
& \qquad - \frac{\partial F}{\partial r}(x, y, z_0, p_0, q_0, r_0, s_0, t_0)\biggr\}\, d\lambda \\
& + (s - s_0)\int_0^1 \biggl\{\frac{\partial F}{\partial s}[x, y, z, p, q, r_0 + \lambda(r - r_0), \\
& \qquad s_0 + \lambda(s - s_0), t_0 + \lambda(t - t_0)] \\
& \qquad - \frac{\partial F}{\partial s}(x, y, z_0, p_0, q_0, r_0, s_0, t_0)\biggr\}\, d\lambda \\
& + (t - t_0)\int_0^1 \biggl\{\frac{\partial F}{\partial t}[x, y, z, p, q, r_0 + \lambda(r - r_0), \\
& \qquad s_0 + \lambda(s - s_0), t_0 + \lambda(t - t_0)] \\
& \qquad - \frac{\partial F}{\partial t}(x, y, z_0, p_0, r_0, s_0, t_0)\biggr\}\, d\lambda
\end{aligned} \tag{10.2}$$

where p_0, q_0, r_0, s_0, t_0 denote derivatives of z_0.

Our next step consists, roughly speaking, in applying the Schauder Existence Theorem (6.2) to solve the linear part of this equation. We write:

$$(10.3)\quad [F_r(x, y, z_0, p_0, q_0, r_0, s_0, t_0)]w_{xx} + [F_s]w_{xy} + [F_t]w_{yy} = \zeta(x, y)$$

where F_s and F_t have the same arguments as F_r and

$$\zeta(x, y) \in C_\alpha(\overline{\mathscr{D}}).$$

Let

$$\phi(s) \in c_{2+\alpha}(\mathscr{D}').$$

Applying the Schauder Existence Theorem to equation (10.3), we obtain the solution

$$w(\zeta, \phi) \in C_{2+\alpha}(\overline{\mathscr{D}}).$$

Now w may be regarded as an operator which takes

$$(\zeta, \phi) \in [C_\alpha(\overline{\mathscr{D}})] \times [c_{2+\alpha}(\mathscr{D}')]$$

into $C_{2+\alpha}(\overline{\mathscr{D}})$, and operator w is linear and 1-1 in ζ and ϕ. Using $w(\zeta, \phi)$, we may rewrite (10.2) as

$$\begin{aligned}
\psi(x, y) = F\Big[&x, y, z_0 + w(\zeta, \phi), p_0 + \frac{\partial}{\partial x} w(\zeta, \phi), \\
&q_0 + \frac{\partial}{\partial y} w(\zeta, \phi), r_0, s_0, t_0\Big] + \zeta(x, y) \\
&+ \frac{\partial^2 w(\zeta, \phi)}{\partial x^2} \int_0^1 \Big\{ \frac{\partial F}{\partial r}\Big[x, y, z_0 + w(\zeta, \phi), p_0 + \frac{\partial}{\partial x} w(\zeta, \phi), \\
&\qquad q_0 + \frac{\partial}{\partial y} w(\zeta, \phi), r_0 + \lambda \frac{\partial^2 w}{\partial x^2}(\zeta, \phi), \\
&\qquad s_0 + \lambda \frac{\partial^2 w}{\partial x\, \partial y}(\zeta, \phi), t_0 + \lambda \frac{\partial^2 w}{\partial y^2}(\zeta, \phi)\Big] \\
(10.4)\qquad &\qquad - \frac{\partial F}{\partial r}[x, y, z_0, p_0, q_0, r_0, s_0, t_0]\Big\} d\lambda \\
&+ \frac{\partial^2 w(\zeta, \phi)}{\partial x\, \partial y} \int_0^1 \Big\{ \frac{\partial F}{\partial s}\Big[x, y, z_0 + w, p_0 + \frac{\partial}{\partial x} w, \\
&\qquad q_0 + \frac{\partial}{\partial y} w, r_0 + \lambda \frac{\partial^2}{\partial x^2} w, \\
&\qquad s_0 + \lambda \frac{\partial^2}{\partial x\, \partial y} w, t_0 + \lambda \frac{\partial^2}{\partial y^2} w\Big] \\
&\qquad - \frac{\partial F}{\partial s}[x, y, z_0, p_0, q_0, r_0, s_0, t_0]\Big\} d\lambda
\end{aligned}$$

$$+ \frac{\partial^2 w}{\partial y^2} \int_0^1 \left\{ \frac{\partial F}{\partial t} \left[x, y, z_0 + w, p_0 + \frac{\partial}{\partial x} w, q_0 + \frac{\partial}{\partial y} w, \right.\right.$$

$$\left. r_0 + \lambda \frac{\partial^2}{\partial x^2} w, s_0 + \lambda \frac{\partial^2}{\partial x \, \partial y} w, t_0 + \lambda \frac{\partial y^2}{\partial^2 w} w \right]$$

$$\left. - \frac{\partial F}{\partial t} [x, y, z_0, p_0, q_0, r_0, s_0, t_0] \right\} d\lambda$$

where ϕ is the given boundary value of the sought-for solution and $\psi(x, y)$ is also given.

We define the transformation from $[C_\alpha(\overline{\mathscr{D}})] \times [c_{2+\alpha}(\mathscr{D}')]$ into $C_\alpha(\overline{\mathscr{D}})$:

$$R^{(1)}(\zeta, \phi) = F\left[x, y, z_0 + w, p_0 + \frac{\partial}{\partial x} w, q_0 + \frac{\partial}{\partial y} w, r_0, s_0, t_0\right].$$

Since $F(x, y, z_0, p_0, q_0, r_0, s_0, t_0) \equiv 0$, then by using Taylor's expansion, $R^{(1)}(\zeta, \phi)$ may be written in the form:

$$R^{(1)}(\zeta, \phi) = R_1^{(1)}(\zeta, \phi) + R_2^{(1)}(\zeta, \phi)$$

where

$$R_1^{(1)}(\zeta, \phi) = [w(\zeta, \phi)] \frac{\partial F}{\partial z} + \left[\frac{\partial}{\partial x} w\right] \left[\frac{\partial F}{\partial p}\right] + \left[\frac{\partial}{\partial y} w\right] \left[\frac{\partial F}{\partial q}\right]$$

in which $\partial F/\partial x$, $\partial F/\partial p$, $\partial F/\partial q$ all have arguments $(x, y, z_0, p_0, q_0, r_0, s_0, t_0)$ and

$$R_2^{(1)}(\zeta, \phi) = [w(\zeta, \phi)]^2 \int_0^1 \left[\frac{\partial^2 F}{\partial z^2} \left(x, y, z_0 + \tau w, p_0 + \tau \frac{\partial}{\partial x} w, \right.\right.$$

$$\left.\left. q_0 + \tau \frac{\partial}{\partial y} w, r_0, s_0, t_0\right)\right] [1 - \tau] \, d\tau$$

$$+ 2(w) \left(\frac{\partial w}{\partial x}\right) \int_0^1 \left(\frac{\partial^2 F}{\partial x \, \partial p}\right) (1 - \tau) \, d\tau$$

$$+ 2(w) \left(\frac{\partial w}{\partial y}\right) \int_0^1 \left(\frac{\partial^2 F}{\partial x \, \partial q}\right) (1 - \tau) \, d\tau$$

$$+ \left(\frac{\partial w}{\partial x}\right)^2 \int_0^1 \left(\frac{\partial^2 F}{\partial p^2}\right) (1 - \tau) \, d\tau$$

$$+ 2 \left(\frac{\partial w}{\partial x}\right) \left(\frac{\partial w}{\partial y}\right) \int_0^1 \left(\frac{\partial^2 F}{\partial p \, \partial q}\right) (1 - \tau) \, d\tau$$

$$+ \left(\frac{\partial w}{\partial y}\right)^2 \int_0^1 \left(\frac{\partial^2 F}{\partial q^2}\right) (1 - \tau) \, d\tau.$$

We define also the transformation

$$R^{(2)}(\zeta, \phi) = \frac{\partial^2 w}{\partial x^2} \int_0^1 \left\{ \frac{\partial F}{\partial r} \left[x, y, z_0 + w, p_0 + \frac{\partial}{\partial x} w, q_0 + \frac{\partial}{\partial y} w, \right.\right.$$
$$\left. r_0 + \lambda \frac{\partial^2}{\partial x^2} w, s_0 + \lambda \frac{\partial^2}{\partial x\, \partial y} w, t_0 + \lambda \frac{\partial^2}{\partial y^2} w \right]$$
$$\left. - \frac{\partial F}{\partial r} (x, y, z_0, p_0, q_0, r_0, s_0, t_0) \right\} d\lambda$$
$$+ \frac{\partial^2 w}{\partial x\, \partial y} \int_0^1 \left\{ \frac{\partial F}{\partial s} \left[x, y, z_0 + w, p_0 + \frac{\partial}{\partial x} w, q_0 + \frac{\partial}{\partial y} w, \right.\right.$$
$$\left. r_0 + \lambda \frac{\partial^2}{\partial x^2} w, s_0 + \lambda \frac{\partial^2}{\partial x\, \partial y} w, t_0 + \lambda \frac{\partial^2}{\partial y^2} w \right]$$
$$\left. - \frac{\partial F}{\partial s} (x, y, z_0, p_0, q_0, r_0, s_0, t_0) \right\} d\lambda$$
$$+ \frac{\partial^2 w}{\partial y^2} \int_0^1 \left\{ \frac{\partial F}{\partial t} \left[x, y, z_0 + w, p_0 + \frac{\partial}{\partial x} w, q_0 + \frac{\partial}{\partial y} w, \right.\right.$$
$$\left. r_0 + \lambda \frac{\partial}{\partial x^2} w, s_0 + \lambda \frac{\partial^2}{\partial x\, \partial y} w, t_0 + \lambda \frac{\partial^2}{\partial y^2} w \right]$$
$$\left. - \frac{\partial F}{\partial t} (x, y, z_0, p_0, q_0, r_0, s_0, t_0) \right\} d\lambda.$$

Now equation (10.4) may be written:

$$\psi = \zeta + R_1^{(1)}(\zeta, \phi) + R_2^{(1)}(\zeta, \phi) + R^{(2)}(\zeta, \phi).$$

Instead of studying this equation, we study the equation in $[C_\alpha(\overline{\mathscr{D}})] \times [c_{2+\alpha}(\mathscr{D}')]$:

$$(\zeta, \phi) + (R_1^{(1)}(\zeta, \phi), \overline{0}) + (R_2^{(1)}(\zeta, \phi), \overline{0}) + (R^{(2)}(\zeta, \phi), \overline{0}) = (\psi, \phi), \tag{10.5}$$

where $\overline{0}$ is the zero element of $c_{2+\alpha}(\mathscr{D}')$. Denoting $(R_1^{(1)}(\zeta, \phi), \overline{0})$ by $C(\zeta, \phi)$ and $(R_2^{(1)}(\zeta, \phi) + R^{(2)}(\zeta, \phi), \overline{0})$ by $T(\zeta, \phi)$ where C and T are transformations from $[C_\alpha(\overline{\mathscr{D}})] \times [c_{2+\alpha}(\mathscr{D}')]$ into itself, we may write (10.5) as:

$$I(\zeta, \phi) + C(\zeta, \phi) + T(\zeta, \phi) = (\psi, \phi) \tag{10.6}$$

where I is the identity transformation.

From the derivation of (10.6), we have: corresponding to each solution (ζ_1, ϕ_1) of (10.6) for given $(\psi_1, \phi_1) \in [C_\alpha(\overline{\mathscr{D}})] \times [c_{2+\alpha}(\mathscr{D}')]$ there is exactly one solution $z_0 + w(\zeta_1, \phi_1)$ of (10.1) with $\psi = \psi_1$ and with boundary value ϕ_1. Since w is a 1-1 operator, this correspondence is 1-1. Hence the original problem is reduced to that of studying (10.6). As proved by Cronin [**2**], transformation C in equation (10.6) is linear and completely continuous and transformation T satisfies conditions (i), (ii), (iii), (iv), and (v) given in

Chapter III. Hence the theory of Schmidt operators can be applied to equation (10.6).

REMARK. The dimension of the null space of the operator $I(\zeta, \phi) + C(\zeta, \phi)$ is equal to the number of linearly independent solutions of the Jacobian equation associated with equation (10.1).

PARABOLIC DIFFERENTIAL EQUATIONS

11. **An analog of the Schauder Existence Theorem.** Studies of parabolic equations analogous to the studies of elliptic equations described in §§ 4–9 of this chapter have been made by several writers (especially Pogorzelski, Barrar, Friedman, and Oleĭnik) and although the results are not as extensive as for elliptic equations, a number of existence theorems have been obtained. The first step is to formulate the problem in terms of a compact mapping. For the general case this means obtaining an analog of the Schauder Existence Theorem (6.2) for parabolic equations. Such a theorem has been obtained by Barrar [**1**; **3**] and Friedman [**1**]. We will describe the theorem for the case in which there are two space variables and the region over which the equation is studied is a particularly simple one. The theorem holds for n space variables and more general types of regions.

Consider the parabolic equation:

$$(11.1) \qquad a_{11}u_{xx} + a_{12}u_{xy} + a_{22}u_{yy} - u_t + b_1u_x + b_2u_y + cu = f$$

in the set

$$\bar{\mathscr{D}} = \bar{D} \times [0, T],$$

where $\bar{D}$ is the closure of a bounded connected open set D in the xy-plane and a_{ij}, b_i $(i, j = 1, 2)$, c and f are functions with domain $\bar{\mathscr{D}}$. We study the Dirichlet problem for (11.1) on $\bar{\mathscr{D}}$, i.e., the problem of finding a solution of (11.1) in $\mathscr{D} = D \times (0, T)$ which has continuous second derivatives in $\mathscr{D}$ and which continuously approaches a given boundary value ϕ on the set $\{D \times [0]\} \cup \{D' \times [0, T]\}$ in the (x, y, t)-space where $D \times [0] = \{(x, y, 0) / (x, y) \in D\}$. (The prototype of this problem is the problem of temperature in a slab with given initial temperature and with faces at given temperatures. Notice that the boundary value ϕ is not specified on the entire boundary $\mathscr{D}'$ of $\mathscr{D}$ as in the Dirichlet problem for elliptic equations.)

Let $p = (x_1, y_1, t_1)$ and $q = (x_2, y_2, t_2)$ be a pair of points in $\mathscr{D}$ and define

$$\bar{d}(p, q) = \sqrt{(x_1 - x_2)^2 + (y_1 - y_2)^2 + |t_1 - t_2|}.$$

Let f be a function with domain $\mathscr{D}$; let

$$H_\alpha(f) = \operatorname*{lub}_{p,\, q \in \mathscr{D}} \frac{|f(p) - f(q)|}{(\bar{d}(p, q))^\alpha}$$

and let

$$\|f\|_\alpha^P = \operatorname*{lub}_{p \in \mathscr{D}} |f(p)| + H_\alpha(f).$$

If f has second derivatives, define:

$$\|f\|^P_{2+\alpha} = \|f\|^P_\alpha + \|f_x\|^P_{(1+\alpha)/2} + \|f_y\|^P_{(1+\alpha)/2} + \|f_t\|^P_\alpha + \|f_{xx}\|^P_\alpha + \|f_{xy}\|^P_\alpha + \|f_{yy}\|^P_\alpha.$$

If $\|f\|^P_{2+\alpha} < \infty$, function f is said to have a $(2 + \alpha)$-norm on $\mathscr{D}$. Notice the similarities and the differences between the norm $\|f\|_{m+\alpha}$ defined in § 5 for use in elliptic equations and this norm. By the same proof as for Lemma (5.1), if $\|f\|^P_{2+\alpha} < \infty$ then f has a continuous extension on $\overline{\mathscr{D}}$.

Let ϕ be a function whose domain is the set:

$$\mathfrak{S} = (D \times [0]) \cup (D' \times [0, T]).$$

Then we can analogously define $\|\phi\|^P_{2+\alpha}$.

A set D in the xy-plane is said to be of class $B_{2+\alpha}$ if for each point $(x, y) \in D'$, there is a neighborhood M of (x, y) and a neighborhood N of the origin in another plane (with coordinates s_1, s_2) and a 1-1 mapping from N onto M described by a pair of real-valued functions $x(s_1, s_2)$ and $y(s_1, s_2)$ such that

$$x(s_1, s_2) \in C_{2+\alpha}(\overline{N}),$$
$$y(s_1, s_2) \in C_{2+\alpha}(\overline{N}),$$

and the boundary points of D are given by

$$(x(s_1, 0), y(s_1, 0)).$$

Finally, the set $\mathscr{D}$ is said to be of class $B_{2+\alpha}$ if D is of class $B_{2+\alpha}$.

(11.2) BARRAR-FRIEDMAN EXISTENCE THEOREM. *Suppose there exist constants α and ε such that*

$$0 < \alpha < \alpha + \varepsilon < 1$$

and such that $\mathscr{D}$ is of type $B_{2+\alpha+\varepsilon}$ and in equation (11.1), *the norms $\|a_i\|^P_{\alpha+\varepsilon}$, $\|b_i\|^P_\alpha$ $(i, j = 1, 2)$ and $\|c\|^P_\alpha$ and $\|f\|^{P'}_\alpha$ (all norms on $\mathscr{D}$) are all less than a positive constant M. Suppose also that the matrix*

$$(a_{ij}(x, y, t))$$

is symmetric and its quadratic form is positive definite on $\mathscr{D}$ which implies there is a positive constant N such that

$$\det (a_{ij}(x, y, t)) \geqq N.$$

Suppose function ϕ with domain $\mathfrak{S}$ and such that

$$\|\phi\|^P_{2+\alpha} < \infty$$

is given and suppose ϕ is compatible with (11.1), *i.e., ϕ satisfies* (11.1) *on $(D' \times [0]) \cup (D' \times [T])$. Then* (11.1) *has a unique solution $u(x, y, t)$ with a $(2 + \alpha)$-norm on $\mathscr{D}$ and*

$$\|u\|^P_{2+\alpha} < C(M, N, \overline{\mathscr{D}})[\|f\|^P_\alpha + \|\phi\|^P_{2+\alpha}],$$

where $C(M, N, \overline{\mathscr{D}})$ is a positive constant which depends only on M, N, and $\overline{\mathscr{D}}$.

Barrar's proof (Barrar [**1**; **3**]) is modelled on the proof of the Schauder Existence Theorem (6.2) for elliptic equations. Friedman [**1**] has given a simpler proof which is analogous to the method for elliptic equations developed by Douglis and Nirenberg [**1**]. Friedman [**3**; **4**] also obtained an a priori $(1 + \delta)$-estimate under weaker conditions on the coefficients in (11.1) and interior estimates for solutions of general parabolic systems.

The uniqueness of the solution in Theorem (11.2) was proved by Gevrey [**1**, paragraph 18]. (The proof of uniqueness depends on the fact that in equation (11.1) the coefficient of $\partial u/\partial t$ is negative. If this condition does not hold, the proof breaks down unless a condition is put on c, the coefficient of u in (11.1).) This uniqueness marks an essential difference between the elliptic and parabolic equations. Even in the most general form of the Schauder Existence Theorem (6.2) in which a term of the form $f(x, y)z$ appears, it is required that $f(x, y) \leqq 0$ for all (x, y). This is because the elliptic equation does not, in general, have a unique solution if $f(x, y) \geqq 0$. This nonuniqueness appears in the simplest kind of elliptic equations, e.g., the eigenvalue problem for the equation:

$$z_{xx} + z_{yy} + \lambda z = 0.$$

See Courant and Hilbert [**1**, Chapter V, especially § 5].

Thus if the theory of Schmidt operators (§ 8 of Chapter III) is applied to a nonlinear parabolic equation, then the Implicit Function Theorem (8.4) of Chapter III yields the existence of a unique local solution for the equation. The more complicated case in which the null space of the operator $I + C$ is different from $\bar{0}$ cannot occur. This is in contrast with elliptic equations where the nonuniqueness of solutions of the linear equation gives rise to an operator $I + C$ with a nonzero null space. (See the Remark at the end of § 10 of this chapter.)

12. **Some results for quasilinear parabolic equations.** In order to obtain a result analogous to the Leray-Schauder-Nirenberg result for elliptic equations, the next step is to obtain either an a priori bound for the solutions of a nonlinear parabolic equation (an analog of Theorem (7.12) obtained by Leray and Schauder) or show that the appropriate mapping is an into mapping (an analog of Theorems (8.4) and (8.6) obtained by Nirenberg).

Oleĭnik [**1**] has announced very general results of this kind for the Cauchy problem and the Dirichlet problem for quasilinear parabolic equations. The following equation is considered:

$$\sum_{i,j=1}^{n} \frac{\partial}{\partial x_i}\left(a_{ij}(t, x, u)\frac{\partial u}{\partial x_j}\right) + A(t, x, u, u_x) - \frac{\partial u}{\partial t} = 0 \tag{12.1}$$

where it is required that for all $x, t \in [0, T]$, $|u| \leqq M$, a positive number, and $[\sum_{i=1}^{n} |\alpha_i|^2]^{1/2} = 1$, we have:

$$0 < \lambda_1(M) \leqq \sum_{i,j=1}^{n} a_{ij}\alpha_i\alpha_j \leqq \lambda_2(M).$$

Also the coefficients a_{ij} and A are sufficiently smooth; if $|u| \leqq M$ and p, x are arbitrary and $t \geqq 0$, functions $A(t, x, u, p)$ and $a_{ij}(t, x, u)$ $(i, j = 1, \cdots, n)$ are bounded; also $a_{ij} = a_{ji}$ and $A_u < c$, a negative constant. The Cauchy problem for (12.1) is the following: if $u_0(x)$ is a bounded smooth function on R^n, to find a solution $u(t, x)$ of (12.1) in $[0, T] \times R^n$ such that

$$u(0, x) = u_0(x). \tag{12.2}$$

Oleĭnik shows: if $u_0(x)$ has continuous third derivatives in R^n and $\lim_{x\to\infty} u_0(x) = 0$; and if for $(t, x) \in [0, T] \times R^n$ and $|u| \leqq M_1$ (where M_1 depends on u_0 and $a(t, x, u, 0)$) the functions a_{ij} have α-Hölder continuous third derivatives and function A has α-Hölder continuous second derivatives and $A(t, x, 0, 0) = 0$, then there exists in $[0, T] \times R^n$ a unique solution $u(t, x)$ of (12.1) satisfying condition (12.2) and such that $u, u_{x_i}, u_t, u_{x_i x_j}$ are bounded in $[0, T] \times R^n$ and satisfy Hölder conditions.

A similar existence and uniqueness theorem is obtained for the solution of the Dirichlet problem for (12.1).

The proofs are obtained by deriving a priori estimates for the solutions (these derivations use results of Nash [**1**]) and then applying the Leray-Schauder technique.

The Schauder Fixed Point Theorem has also been applied to the study of a more restricted class of quasilinear parabolic equations, especially by Pogorzelski [**2–10**] and Friedman [**1–3**] and also by Prodi [**1**], Piskorek [**1**], and Malferrai [**1**]. We describe roughly a few typical results.

Pogorzelski [**4**; **5**] has constructed a fundamental solution for parabolic equations with coefficients which are Hölder continuous functions of the n space variables $x_1, \cdots, x_n$ and the time variable t. Using this result of Pogorzelski, Friedman [**3**] has established the existence and uniqueness of a solution of the first mixed boundary problem for a parabolic equation:

$$\sum_{i,j=1}^{n} a_{ij}(x, t) \frac{\partial^2 u}{\partial x_i \, \partial x_j} + \sum_{i=1}^{n} b_i(x, t) \frac{\partial u}{\partial x_i} + c(x, t)u - \frac{\partial u}{\partial t} = f(x, t, u, \nabla u) \quad \text{for } (x, t) \in \mathscr{D}, \tag{12.3}$$

$$u = h(x, t) \quad \text{for } (x, t) \in \mathscr{D}'$$

where $\mathscr{D}$ is a connected open set in Euclidean $(n + 1)$-space such that $\mathscr{D}'$, the boundary of $\mathscr{D}$, consists of subsets of the hyperplanes $t = 0$ and $t = t_0 > 0$ and a surface S between the hyperplanes; and

$$(x, t) = (x_1, \cdots, x_n, t),$$

$$\nabla u = \left(\frac{\partial u}{\partial x_i}, \cdots, \frac{\partial u}{\partial x_n}\right),$$

and $h(x, t)$ is a given function. It is assumed that the coefficients satisfy Hölder continuity conditions and that f satisfies the conditions: there exist positive constants A, B, C, with C sufficiently small, such that: for $(x, t) \in \overline{\mathscr{D}}$, and $-\infty < u < \infty$, and $-\infty < \partial u/\partial x_i < \infty$ $(i = 1, \cdots, n)$,

$$|f(x, t, u, \nabla u)| \leqq A + B|u| + C|\nabla u|$$

where $|\nabla u| = \sum_{i=1}^{n} |\partial u/\partial x_i|$.

Special cases of this result had previously been obtained by Pogorzelski (for example, Pogorzelski [**3**]); and using his own results [**4**; **5**], Pogorzelski [**7**] solved the second mixed boundary problem for equation (12.3). Pogorzelski [**10**] has also studied a case in which the coefficients are functions of the unknown u. He considers the parabolic equation:

$$\sum_{i,j=1}^{n} a_{ij} \frac{\partial^2 u}{\partial x_i \, \partial x_j} + \sum_{i=1}^{n} b_i \frac{\partial u}{\partial x_i} + cu - \frac{\partial u}{\partial t} = F$$

where a_{ij}, b_i, c are Hölder continuous in $x_1, \cdots, x_n$, t, u and F is Hölder continuous in $x_1, \cdots, x_n$, t, u, $\partial u/\partial x_1, \cdots, \partial u/\partial x_n$. However, the Hölder constants with respect to u, $\partial u/\partial x_1, \cdots, \partial u/\partial x_n$ of the functions are assumed to be very small.

Hyperbolic Differential Equations

13. **The Cauchy problem.** Schauder [**5**] has established the existence of a solution of the Cauchy problem for a quasilinear hyperbolic equation by using the Schauder Fixed Point Theorem. The solution obtained is not necessarily analytic; only differentiability conditions of sufficiently higher order are required of the solution. However, as Schauder points out [**5**, p. 219, footnote 17 and p. 245, footnote 46] the solution is unique and can be obtained by successive approximations. Although a topological treatment is interesting, it is therefore unnecessary. In keeping with the discussion in the Foreword, we say that this is a case of when not to use a topological approach.

14. **Mixed problems.** The mixed initial and boundary-value problem has been studied by Krzyzanski and Schauder [**1**] for quasilinear hyperbolic equations and by Schauder [**6**] for general nonlinear hyperbolic equations. Other problems for nonlinear hyperbolic equations, including the Goursat problem and the Darboux problem, have been studied by Ciliberto [**1**], Szmydt [**1–7**] and Guiglielmo [**1**]. In all of these papers, the principal tool is the Schauder Fixed Point Theorem. In some of the results, the solution obtained is unique which suggests the possibility of an analytic approach. However, in other problems the solution is not unique or the uniqueness question has not been settled.

Ordinary Differential Equations

15. **The Cesari method.** Finally we describe briefly a topological method for studying periodic solutions of systems of ordinary nonlinear differential

equations recently developed by Cesari [**1**]. This method, which uses the Schauder or Banach Fixed Point Theorems and the local degree of mappings in Euclidean n-space, goes considerably beyond the techniques and results described for nonlinear systems in § 9 of Chapter II. Unlike the results of Chapter II which are, for the most part, only applicable to 2-dimensional systems, Cesari's method is applicable to a system of arbitrary dimension. Also Cesari's method can be used to obtain approximations to the periodic solution while the results in Chapter II yield no information about numerical values of the solutions. Of course, a price must be paid for these advantages: the computations in Cesari's method are lengthy and thus far have been carried out completely (i.e., through the determination of the degree) only for comparatively simple systems. But it should be emphasized that even before such complete computations are carried out, Cesari's method sheds considerable light on the nature of the problem. Moreover, all previous work on the question of periodic solutions of nonlinear systems indicates that it is a very difficult problem and we could hardly expect to obtain a simple method for solving it.

The Cesari method has also been recently used (Cesari [**4**; **5**]) to study boundary value problems for nonlinear ordinary differential equations and nonlinear partial differential equations. In particular, striking results for nonlinear partial differential equations have been obtained (Cesari [**5**]).

Consider an n-dimensional system

$$\dot{x}_j = q_j(x_1, \cdots, x_n, t) \quad (j = 1, \cdots, n), \tag{15.1}$$

where each $q_j(x_1, \cdots, x_n, t)$ has period $\tau = 2\pi/w$ in t. The problem is to determine if system (15.1) has a solution of period τ. (The technique can also be used to study periodic solutions in the autonomous case.) We assume that there exist functions $K(t)$, defined for all real t and square-integrable on $[0, \tau]$, and $\eta(\xi)$, $\xi \geqq 0$, continuous and monotonic nondecreasing with $\eta(0) = 0$, such that for $|x_j| \leqq R_j$, $|x_j^1| \leqq R_j$ and $|x_j^2| \leqq R_j$ where the R_j are positive constants for $j = 1, \cdots, n$, and $-\infty < t < \infty$,

$$\left.\begin{aligned} q_j(x, t) &\leqq K(t), \\ |q_j(x^1, t) - q_j(x^2, t)| &\leqq \eta(\xi)K(t) \end{aligned}\right. \quad j = 1, \cdots, n,$$

where $\xi = \max_{s=1,\cdots,n} |x_s^1 - x_s^2|$.

Let S be the linear space of real vector functions

$$x(t) = (x_1(t), \cdots, x_n(t))$$

where for each j, function $x_j(t)$ has period τ and

$$\int_0^\tau [x_j(t)]^2 \, dt < \infty.$$

Let

$$v(x_j) = \left[\frac{1}{\tau}\int_0^\tau [x_j(t)]^2\,dt\right]^{1/2}.$$

If

$$x_j(t) \sim a_{j0} + \sum_{s=1}^{\infty} (a_{js}\cos swt + b_{js}\sin swt),$$

then

$$\nu(x_j) = a_{j0}^2 + \frac{1}{2}\sum_{j=1}^{\infty} (a_{js}^2 + b_{js}^2)^{1/2}.$$

Then

$$\nu(x) = \max_j \nu(x_j)$$

is a norm for S (as usual, functions which differ only on a set of measure zero are regarded as identical). For m a positive integer, define the operator

$$P(x) = (P_1x_1, \cdots, P_nx_n)$$

where

$$P_jx_j = a_{j0} + \sum_{s=1}^{m} (a_{js}\cos swt + b_{js}\sin swt).$$

Let

$$\tilde{S} = [x \,/\, x \in S \text{ and } Px = 0].$$

If $x \in \tilde{S}$, then

$$x_j(t) \sim \sum_{s=m+1}^{\infty} (a_{js}\cos swt + b_{js}\sin swt).$$

For $x \in \tilde{S}$, we define an operator H such that

$$H(x) = (X_1, \cdots, X_n)$$

where

$$X_j(t) = \sum_{s=m+1}^{\infty} \frac{1}{sw}(-b_{js}\cos swt + a_{js}\sin swt).$$

Then H is a linear operator such that

$$H(\tilde{S}) \subset \tilde{S}$$

and for $x \in \tilde{S}$,

$$\nu(H(x)) \leqq \frac{1}{(m+1)w}\nu(x).$$

If $x \in S$, then $x - P(x) \in \tilde{S}$ and hence

$$\nu[H(x - Px)] \leqq \frac{1}{(m+1)w}\nu(x - Px) \leqq \frac{1}{(m+1)w}\nu(x).$$

Now let

$$S_R = [x \in S \,/\, |x_j| \leqq R_j, j = 1, \cdots, n],$$

and define operators q, f, F in S_R as follows:

$$q(x) = (q_1 x, \cdots, q_n x)$$

where

$$q_j x = q_j[x(t), t] \qquad (j = 1, \cdots, n);$$

and

$$f(x) = (q - Pq)(x);$$

$$F(x) = Hf(x) = H(q - Pq)(x).$$

Now let

$$K_0 = \left[\tau^{-1} \int_0^\tau [K(t)]^2 \, dt\right]^{1/2}.$$

Then Cesari [**1**, pp. 12–14] proves that for $x \in S_R$,

$$\nu(F(x)) \leqq K_0(m + 1)^{-1} w^{-1}$$

and (p. 15)

$$|F_j x| \leqq \sqrt{2}\, K_0 w^{-1} \sigma(m)$$

where $[\sigma(m)]^2 = (m + 1)^{-2} + (m + 2)^{-2} + \cdots$.

Now let c and r_j be positive constants where $r_j < R_j$ $(j = 1, \cdots, n)$ and let x^* be an element of S such that

$$\nu(Px^*) \leqq c$$

and

$$|Px_j^*| < r_j$$

for all t. Let d be a constant such that $c < d$ and define

$$S_R^* = [x \,/\, x \in S, \; Px = Px^*, \; \nu(x) \leqq d, \; |x_j| \leqq R_j \quad \text{for } j = 1, \cdots, n].$$

Let m be large enough so that

$$K_0(m + 1)^{-1} w^{-1} \leqq d - c$$

and

$$\sqrt{2}\, K_0 \sigma(m) w^{-1} \leqq R_j - r_j.$$

Then it follows that the transformation

$$T = P + F$$

is a continuous mapping of S_R^* into itself and that T/S_R^* is compact. Hence by the Schauder Fixed Point Theorem there is at least one point y of S_R^* which is taken into itself by T. If

$$\eta(\xi) = \xi, \qquad K(t) = M,$$

where M is a positive constant, then T is a contraction mapping and by the Banach Fixed Point Theorem, transformation T has a unique fixed point $y \in S_R^*$. Moreover y is a continuous function $y[x^*]$ of x^*.

Since y is a fixed point, then

$$y = Py + Fy$$

and for $j = 1, \cdots, n$,

$$\dot{y}_j = P\dot{y}_j + \frac{d}{dt} F_j y = P\dot{y}_j + q_j(y(t), t) - Pq_j(y(t), t)$$

almost everywhere. Thus

$$\dot{y}_j = q_j(y(t), t) + P_j(\dot{y}_j - q_j y).$$

Hence $y(t)$ is a solution of (15.1) of period τ if

$$P_j(\dot{y}_j - q_j(y)) = 0 \tag{15.2}$$

for $j = 1, \cdots, n$, or

$$\tau^{-1} \int_0^\tau [\dot{y}_j - q_j(y(t), t)] \begin{Bmatrix} 1 \\ \cos swt \\ \sin swt \end{Bmatrix} dt = 0 \tag{15.3}$$

for $s = 1, \cdots, m$ and $j = 1, \cdots, n$.

Setting

$$y_j(t) = a_{j0} + \sum_{s=1}^{\infty} (a_{js} \cos swt + b_{js} \sin swt),$$

$$q_j(y(t), t) = A_{j0} + \sum_{s=1}^{\infty} (A_{js} \cos swt + B_{js} \sin swt)$$

we can write (15.2) as

$$A_{j0} + \sum_{s=1}^{m} [(swb_{js} + A_{js}) \cos swt + (-swa_{js} + B_{js}) \sin swt] = 0.$$

Then (15.3) becomes:

$$\begin{aligned} A_{j0} &= 0, \qquad j = 1, \cdots, n, \\ swb_{js} + A_{js} &= 0, \qquad s = 1, \cdots, m, \qquad j = 1, \cdots, n, \\ -swa_{js} + B_{js} &= 0, \qquad s = 1, \cdots, m, \qquad j = 1, \cdots, n. \end{aligned} \tag{15.4}$$

Now each A_{js} and B_{js} is a function of the a_{js} and b_{js}, the coefficients in the Fourier expansion of $y_j(t)$. Hence if x^* can be so chosen that the a_{js}, b_{js} in $y(x^*)$ satisfy the system (15.4), then $y[x^*]$ is the desired periodic solution of

(15.1). System (15.4) is called the system of determining equations. Note especially that if $x^* = Px^*$ so that

$$x_j^* = P_j x^* = a_{j0} + \sum_{s=1}^{m} (a_{js} \cos swt + b_{js} \sin swt)$$

for $j = 1, \cdots, n$ and if T is a contraction so that $y(x^*)$ is a (single-valued) continuous function of x^*, then the coefficients

$$a_{js}, b_{js} \qquad (s \geqq m + 1, j = 1, \cdots, n)$$

are single-valued continuous functions of the coefficients

$$a_{j0}, a_{js}, b_{js} \qquad (s = 1, \cdots, m, j = 1, \cdots, n)$$

because the coefficients in the Fourier expansions of the components of $y[x^*]$ are continuous functions of the coefficients in the Fourier expansions of the components of x^*. Since $x^* = Px^* = Py$, these coefficients are exactly the numbers:

$$a_{j0}, a_{js}, b_{js} \qquad (s = 1, \cdots, m, j = 1, \cdots, n).$$

In this case, (15.4) is a system of $(2m + 1)n$ equations in $(2m + 1)n$ unknowns.

The solutions of system (15.4) are studied by computing the local degree of the mapping from Euclidean $[(2m + 1)n]$-space into itself described by the left sides of (15.4). Cesari describes a general technique for such a study. In equation (15.3) replace $y_j(t)$ by

$$x_j^m(t) = x_j^*(t)$$

where $x^*(t)$ has undetermined coefficients a_{j0}, a_{js}, b_{js} $(s = 1, \cdots, m)$. (The resulting equations are the equations for the mth approximation in the Galerkin method.) It can be shown in some cases that the mapping described by the left sides of the resulting equations has the same local degree as the mapping described by the left sides of (15.4) and also that the local degree of this second mapping can be computed.

This method contains as a particular case the much simpler method previously discussed by Cesari for perturbation problems of any dimension [**2**, pp. 123–136; **3**].

BIBLIOGRAPHY

P. Alexandroff and H. Hopf

1. *Topologie*. I, Springer, Berlin, 1935. (Reprinted by Edwards Brothers, Ann Arbor, Mich., 1945.)

A. Andronov and S. Chaikin

1. *Theory of oscillations*, English language edition, Princeton Univ. Press, Princeton, N. J., 1949.

A. Andronov and A. Witt

1. *Zur Theorie des Mitnehmens von van der Pol*, Arch. Electrotech. **24** (1930), 99–110.

S. Banach

1. *Théorie des opérations linéaires*, Monogr. Mat., Tom I, Warsaw, 1932.

R. Barrar

1. *Some estimates for the solutions of linear parabolic equations*, Thesis, Univ. of Michigan, Ann Arbor, Mich., 1952.

2. *On Schauder's paper on linear elliptic differential equations*, J. Math. Anal. Appl. **3** (1961), 171–195.

3. *Some estimates for solutions of parabolic equations*, J. Math. Anal. Appl. **3** (1961), 373–397.

R. Bartle

1. *Singular points of functional equations*, Trans. Amer. Math. Soc. **75** (1953), 366–384.

S. Bernstein

1. *Sur la généralisation du problème de Dirichlet*. I, Math. Ann. **62** (1906), 253–271.

2. *Sur la généralisation du problème de Dirichlet*. II, Math. Ann. **69** (1910), 82–136.

Lipman Bers

1. *Theory of pseudo-analytic functions*, (mimeographed), New York University, New York, 1953.

G. D. Birkhoff and O. D. Kellogg

1. *Invariant points in function space*, Trans. Amer. Math. Soc. **23** (1922), 96–115.

J. F. Blackburn and W. R. Mann

1. *A nonlinear steady state temperature problem*, Proc. Amer. Math. Soc. **5** (1954), 979–986.

N. N. Bogoliubov and Yu. A. Mitropolski

1. *Asymptotic methods in the theory of nonlinear oscillations*, Gosudarstv. Izdat. Tehn.-Teor. Lit., Moscow, 1955. (Russian) English Transl., Gordon and Breach, New York, 1961.

D. G. Bourgin

1. *Un indice dei punti uniti*. I, II, III, Atti. Accad. Naz. Lincei. Rend. Cl. Sci. Fis. Mat. Nat. (8) **19** (1955), 435–440; **20** (1956), 43–48; **21** (1956), 395–400.

F. Browder

1. *On a generalization of the Schauder fixed point theorem*, Duke Math. J. **26** (1959), 291–303.

2. Thesis, Princeton Univ., Princeton, N. J., 1948.

3. *On continuity of fixed points under deformations of continuous mappings*, Summa Brasil. Math. **4** (1960), Fasc. 5.

4. *On the fixed point index for continuous mappings of locally connected spaces*, Summa Brasil. Math. **4** (1960), Fasc. 7.

R. C. Buck

1. *Advanced calculus*, McGraw-Hill, New York, 1956.

L. Cesari

1. *Functional analysis and periodic solutions of nonlinear differential equations*, MRC Tech. Summary Rep. No. 173, Contract No. DA-11-022-ORD-2059, Math. Research Center, United States Army, Univ. of Wisconsin, Madison, Wis., 1960. (To appear in Contributions to Differential Equations, Vol. 1, Wiley, New York, 1963.)

2. *Asymptotic behavior and stability problems in ordinary differential equations*, Springer, Berlin, 1959.

3. *Existence theorems for periodic solutions of nonlinear Lipschitzian differential systems and fixed point theorems*, Contributions to the theory of nonlinear oscillations, Vol. 5, pp. 115–172, Ann. of Math. Studies No. 45, Princeton Univ. Press, Princeton, N. J., 1960.

4. *Functional analysis and Galerkin's method*, N.S.F. Research Project GP-57, Rep. No. 1, Department of Mathematics, Univ. of Michigan, Ann Arbor, Mich.

5. *A nonlinear problem in potential theory*, N.S.F. Research Project GP-57, Rep. No. 2, Department of Mathematics, Univ. of Michigan, Ann Arbor, Mich., 1962.

C. Ciliberto

1. *Sul problema de Darboux per l'equazione* $s = f(x, y, z, p, q)$, Rend. Accad. Sci. Fis. Mat. Napoli (4) **22** (1955), 221–225 (1956).

E. Coddington and N. Levinson

1. *Theory of ordinary differential equations*, McGraw-Hill, New York, 1955.

H. O. Cordes

1. *Über die erste Randwertaufgabe bei quasilinearen Differentialgleichungen zweiter Ordnung in mehr als zwei Variablen*, Math. Ann. **131** (1956), 278–313.

R. Courant and D. Hilbert

1. *Methods of mathematical physics*, Vol. 1, Interscience, New York, 1953.

2. *Methoden der mathematischen Physik*, Vol. 2, Springer, Berlin, 1937.

J. Cronin

1. *Branch points of solutions of equations in Banach space*, Trans. Amer. Math. Soc. **69** (1950), 208-231.

2. *Branch points of solutions of equations in Banach space*. II, Trans. Amer. Math. Soc. **76** (1954), 207–222.

3. *The existence of multiple solutions of elliptic differential equations*, Trans. Amer. Math. Soc. **68** (1950), 105–131.

4. *The number of periodic solutions of non-autonomous systems*, Duke Math. J. **27** (1960), 183–194.

5. *Poincaré's perturbation method and topological degree*, Contributions to the theory of nonlinear oscillations, Vol. 5, pp. 37–54, Ann. of Math. Studies No. 45, Princeton Univ. Press, Princeton, N. J., 1960.

6. *Stability of periodic solutions of the perturbation problem*, J. Math. Mech. **10** (1961), 19–30.

Mahlon M. Day

1. *Normed linear spaces*, Springer, Berlin, 1958.

A. Douglis and L. Nirenberg

1. *Interior estimates for elliptic systems of partial differential equations*, Comm. Pure Appl. Math. **8** (1955), 503–538.

G. F. D. Duff

1. *Eigenvalues and maximal domains for a quasilinear elliptic equation*, Math. Ann. **131** (1956), 28–37.

R. Finn and D. Gilbarg

1. *Three-dimensional subsonic flows and asymptotic estimates for elliptic partial differential equations*, Acta Math. **98** (1957), 265–296.

A. Friedman

1. *Boundary estimates for second order parabolic equations and their applications*, J. Math. Mech. **7** (1958), 771–792.
2. *On quasi-linear parabolic equations of the second order*, J. Math. Mech. **7** (1958), 793–810.
3. *On quasi-linear parabolic equations of the second order*. II, J. Math. Mech. **9** (1960), 539–556.
4. *Interior estimates for parabolic systems of partial differential equations*, J. Math. Mech. **7** (1958), 393–418.

M. Gevrey

1. *Sur les equations aux derivées partielles du type parabolique*, Journal de Mathématique (6), **9** (1913), 305–471.

R. Gomory

1. *Critical points at infinity and forced oscillations*, Contributions to the theory of nonlinear oscillations, Vol. 3, pp. 85–126, Ann. of Math. Studies No. 36, Princeton Univ. Press, Princeton, N. J., 1956.

L. M. Graves

1. *The estimates of Schauder and their application to existence theorems for elliptic differential equations*, Tech. Rep. No. 1, Contract No. DA-11-022-ORD-1833, University of Chicago, Chicago, Ill., 1956.

F. Guglielmo

1. *Sul problema di Darboux*, Ricerche Math. **8** (1959), 180–196.

J. Hadamard

1. *Lectures on Cauchy's problem in linear partial differential equations*, Dover, New York, 1952.

J. K. Hale

1. *Integral manifolds of perturbed differential systems*, Ann. of Math. **73** (1961), 496–531.

P. Halmos

1. *Introduction to Hilbert space and the theory of spectral multiplicity*, Chelsea, New York, 1951.

E. Heinz

1. *Über die Existenz einer Fläche konstanter mittlere Krümmung bei vorgegebener Berandung*, Math. Ann. **127** (1954), 258–287.

T. H. Hildebrandt and L. M. Growes

1. *Implicit functions and their differentials in general analysis*, Trans. Amer. Math. Soc. **29** (1927), 125–153.

E. Hopf

1. *Elementare Bemerkungen über die Losungen partieller Differentialgleichungen zweiter Ordnung vom elliptischen Typus*, S.-B. Preussischen Akad. Wiss., **19** (1927), 147–152.

W. Hurewicz

1. *Lectures on ordinary differential equations*, Technology Press of the Massachusetts Institute of Technology, Cambridge, Massachusetts, 1958.

P. Jordan and J. von Neumann

1. *On inner products in linear metric spaces*, Ann. of Math. **36** (1935), 719–723.

S. Kakutani

1. *Topological properties of the unit sphere in Hilbert space*, Proc. Imp. Acad. Tokyo **19** (1943), 269–271.
2. *A generalization of Brouwer's fixed-point theorem*, Duke Math. J. **8** (1941), 457–459.

V. L. Klee

1. *Some topological properties of convex sets*, Trans. Amer. Math. Soc. **78** (1955), 30–45.
2. *Shrinkable neighborhoods in Hausdorff linear spaces*, Math. Ann. **141** (1960), 281–285.
3. *Leray-Schauder theory without local convexity*, Math. Ann. **141** (1960), 286–296.
4. *Convex sets in linear spaces*, Duke Math. J. **18** (1951), 443–466.
5. *Convex sets in linear spaces*. II, Duke Math. J. **18** (1951), 875–883.

M. A. Krasnos'elskiĭ

1. *Topological methods in the theory of nonlinear integral equations*, Gosudarstv. Izdat. Tehn. Teor. Lit., Moscow, 1956. (Russian)

M. A. Krasnos'elskiĭ and Ya. B. Rutickiĭ

1. *Orlicz spaces and nonlinear integral equations*, Trudy Moskov. Mat. Obšč. **7** (1958), 63–120. (Russian)

M. Krzyzanski and J. Schauder

1. *Quasilineare Differentialgleichungen zweiter Ordnung vom hyperbolischen Typus: Gemischte Randwertaufgaben*, Studia Math. **6** (1936), 162–189.

Walter T. Kyner

1. *A fixed point theorem*, Contributions to the theory of nonlinear oscillations, Vol. 3, pp. 197–205, Ann. of Math. Studies No. 36, Princeton Univ. Press, Princeton, N. J., 1956.
2. *Small periodic perturbations of an autonomous system of vector equations*, Contributions to the theory of nonlinear oscillations, Vol. 4, pp. 111–125, Ann. of Math. Studies, No. 41, Princeton Univ. Press, Princeton, N. J., 1958.

J. La Salle and S. Lefschetz

1. *Stability by Liapunov's direct method with applications*, Academic Press, New York, 1961.

S. Lefschetz

1. *Existence of periodic solutions for certain differential equations*, Proc. Nat. Acad. Sci. **29** (1943), 29–32.
2. *Algebraic topology*, Amer. Math. Soc. Colloq. Publ. Vol. 27, Amer. Math. Soc., Providence, R. I., 1942.
3. *Topics in topology*, Ann. of Math. Studies No. 10, Princeton Univ. Press, Princeton, N. J., 1942.
4. *Differential equations: geometric theory*, Interscience, New York, 1957, 2nd ed., 1963.

E. Leimanis and N. Minorsky

1. *Dynamics and nonlinear mechanics*, John Wiley, New York, 1958.

J. Leray

1. *Note, Topologie des espaces abstracts de M. Banach*, C. R. Acad. Sci. Paris **200** (1935), 1082–1084.
2. *Les problèmes non linéaires*, Enseignement Math. Nos. 1–2, **30** (1936), 141ff. International Conference on partial differential equations: proper conditions to determine solutions.
3. *Valeurs propres et vecteurs propres d'un endomorphisme complètement continu d'un espace vectoriel à voisinages convexes*, Acta Sci. Math. (Szeged) **12** (1950), 177–186.

J. Leray and J. Schauder

1. *Topologie et équations fonctionelles*, Ann. Sci. École Norm. Sup. No. 3, **51** (1934), 45–78.

N. Levinson

1. *Transformation theory of nonlinear differential equations of the second order*, Ann. of Math. (2), **45** (1944), 723–737.

C. C. MacDuffee

1. *Vectors and matrices*, Carus Mathematical Monographs, No. 7, Mathematical Association of America, 1943.

Edward J. McShane and Truman A. Botts

1. *Real Analysis*, Van Nostrand, New York, 1959.

A. Malferrai

1. *Sur certe equazione quasi lineare di tipo parabolico di ordine superiore al secondo*, Atti. Sem. Mat. Fis. Univ. Modena **8** (1958/59), 174–216.

I. G. Malkin

1. *Some problems in the theory of nonlinear oscillations*, GITTL, Moscow, 1956. (Russian) English Transl., U.S. Atomic Energy Commission Translation AEC-tr-3766, Books 1 and 2.
2. *Theory of stability of motion*, GITTL, Moscow, 1952. (Russian) English Transl., U.S. Atomic Energy Commission Translation AEC-tr-3352.

M. Marden

1. *The geometry of the zeros of a polynomial in a complex variable*, Math. Surveys No. 3, Amer. Math. Soc., Providence, R. I., 1949.

Z. Mikolajska

1. *Sur l'existence de solutions périodique de l'équation différentielle du second ordre dépendant d'un parametre*, Ann. Polon. Math. **6** (1959–60), 51–68.

N. Minorsky

1. *Introduction to non-linear mechanics*, J. W. Edwards, Ann Arbor, Mich., 1947.
2. *Nonlinear oscillations*, Van Nostrand, Princeton, N. J., 1962.

C. Miranda

1. *Equazioni alle derivate parziali di tipo ellittico*, Springer, Berlin, 1955.

M. Nagumo

1. *A theory of degree of mapping based on infinitesimal analysis*, Amer. J. Math. **73** (1951), 485–496.
2. *Degree of mapping in convex linear topological spaces*, Amer. J. Math. **73** (1951), 497–511.

J. Nash

1. *Continuity of solutions of parabolic and elliptic equations*, Amer. J. Math. **80** (1958), 931–954.

L. Nirenberg

1. *On nonlinear elliptic partial differential equations and Hölder continuity*, Comm. Pure Appl. Math. **6** (1953), 103–156.
2. *On a generalization of quasi-conformal mappings and its application to elliptic partial differential equations*, Contributions to the theory of partial differential equations, pp. 95–100. Ann. Math. Studies, No. 33, Princeton Univ. Press, Princeton, N. J., 1954.

Johannes Nitsche

1. *Untersuchungen über die linearen Randwertprobleme linearer und quasilinearer elliptischer Differentialgleichungssysteme.* I, II, Math. Nachr. **14** (1955), 75–127, 157–182.

O. Oleĭnik

1. *Quasi-linear parabolic equations with many independent variables*, Dokl. Akad. Nauk SSSR **138** (1961), 43–46. (Russian) English Transl., Soviet Math. Dokl. **2** (1961), 529–532.

A. Piskorek

1. *Sur certains problèmes aux limites pour l'équation semi-linéare parabolique normale*, Bull. Acad. Polon. Sci. Sér. Sci. Math. Astronom. Phys. **6** (1958), 505–510.

W. Pogorzelski

1. *Sur l'équation intégrale singulière non linéaire et sur les propriétés d'une integrale singulière pour les arcs non fermées*, J. Math. Mech. **7** (1958), 515–532.
2. *Problème aux limites pour l'équation parabolique dont les coefficients dépendent de la fonction inconnue*, Ricerche Mat. **5** (1956), 258–272.
3. *Sur le problème de Fourier généralisé*, Ann. Polon. Math. **3** (1956), 126–141.
4. *Étude de la solution fondamentale de l'équation parabolique*, Ricerche Mat. **5** (1956), 25–57.
5. *Propriétés des integrales de l'équation parabolique normale*, Ann. Polon. Math. **4** (1957), 61–92.
6. *Problèmes aux limites pour l'équation parabolique normale*, Ann. Polon. Math. **4** (1957), 110–126.
7. *Problème aux limites aux derivées tangentielles pour l'équation parabolique*, Ann. Sci. École Norm. Sup. (3) **75** (1958), 19–35.
8. *Étude d'une fonction de Green et du problème aux limites pour l'équation parabolique normale*, Ann. Polon. Math. **4** (1958), 288–307.
9. *Propriétés des integrales généralisées de Poisson-Weierstrass et problème de Cauchy pour un système parabolique*, Ann. Sci. École Norm. Sup. (3) **76** (1959), 125–149.
10. *Premier problème de Fourier pour l'équation parabolique dont les coefficients dépendent de la fonction inconnue*, Ann. Polon. Math. **6** (1959/60), 15–40.

H. Poincaré

1. *Les méthodes nouvelles de la mécanique celeste*, Vols. 1, 2, 3, Gauthiers-Villars, Paris, 1892–99. Reprinted by Dover, New York, 1957.
2. *Sur les courbes défines par les équations differentielles*, Vol. 1, Oeuvres, Gauthier-Villars, Paris, 1892.

G. Prodi

1. *Teoremi di esistenza per equazioni alle derivate parziali non lineári di tipo parabolico.* I, II, Ist. Lombardo Sci. Lett. Rend. Cl. Sci. Mat. Nat. (3) **17** (**86**) (1953), 3–26, 27–47.

T. RADO

1. *Geometrische Betrachtungen über zweidimensionale reguläre Variationsprobleme*, Acta Lit. Sci. Regiae Univ. Hungar. Francisco-Josephine Sectio Sci. Math. (Szeged) (1924–1926), 228–253.

T. RADO and P. V. REICHELDERFER

1. *Continuous transformations in analysis*, Springer, Berlin, 1955.

F. RIESZ

1. *Über lineare Funktionalgleichungen*, Acta Math. **41** (1917), 71–98.

F. RIESZ and B. SZ.-NAGY

1. *Functional analysis*, (translated from the 2nd French edition), Ungar, New York, 1955.

WALTER RUDIN

1. *Principles of mathematical analysis*, McGraw-Hill, New York, 1953.

A. SARD

1. *The measure of the critical values of differentiable maps*, Bull. Amer. Math. Soc. **48** (1942), 883–890.

HELMUT SCHAEFER

1. *Über die Methode der a priori Schranken*, Math. Ann. **129** (1955), 415–416.

J. SCHAUDER

1. *Der Fixpunktsatz in Funktionalraumen*, Studia Math. **2** (1930), 171–180.
2. *Über das Dirichletsche Problem in Grossen für nichtlineare elliptische Differentialgleichungen*, Math. Z. **34** (1933), 623–634.
3. *Über lineare elliptische Differentialgleichungen zweiter Ordnung*, Math. Z. **38** (1934), 257–282.
4. *Numerische Abschatzungen in elliptischen linearen Differentialgleichungen*, Studia Math. **5** (1934), 34–42.
5. *Das Anfangswertproblem einer quasilinearen hyperbolischen Differentialgleichung zweiter Ordnung in beliebiger Anzahl von unabhängigen Veränderlichen*, Fund. Math. **24** (1935), 213–246.
6. *Gemischte Randwertaufgaben bei partiellen Differentialgleichungen vom hyperbolischen Typus*, Studia Math. **6** (1936), 190–198.
7. *Über den Zusammenhang zwischen der Eindeutigkeit und Lösbarkeit partieller Differentialgleichungen zweiter Ordnung von elliptischen Typus*, Math. Ann. **106** (1932), 661–721.

E. SCHMIDT

1. *Zur Theorie der linearen und nichtlinearen Integralgleichungen*, Math. Ann. **65** (1907–1908), 370–399.

J. STOKER

1. *Nonlinear vibrations in mechanical and electrical systems*, Interscience, New York, 1950.

M. H. STONE

1. *Linear transformations in Hilbert space and their applications to analysis*, Amer. Math. Soc. Colloq. Publ. Vol. 15, Amer. Math. Soc., Providence, R. I., 1932.

Z. SZMYDT

1. *Sur une généralisation des problèmes classiques concernant un système d'équations différentielles hyperboliques du second ordre à deux variables indépendantes*, Bull. Acad. Polon. Sci. Sér. Sci. Math. Astronom. Phys. **4** (1956), 579–584.

2. *Sur un nouveau type de problèmes pour un système d'équations différentielles hyperboliques du second ordre à deux variables indépendantes*, Bull. Acad. Polon. Sci. Sér. Sci. Math. Astronom. Phys. **4** (1956), 67–72.
3. *Sur le problème concernant les équations différentielles hyperboliques du second ordre*, Bull. Acad. Polon. Sci. Sér. Sci. Math. Astronom. Phys. **5** (1957), 571–575.
4. *Sur un problème concernant un système d'équations différentielles hyperboliques d'ordre arbitrare à deux variables indépendantes*, Bull. Acad. Polon. Sci. Sér. Sci. Math. Astronom. Phys. **5** (1957), 577–582.
5. *Sur l'existence de solutions de certains problèmes aux limites relatifs a un système d'équations différentielles hyperboliques*, Bull. Acad. Polon. Sci. Sér. Sci. Math., Astronom. Phys. **6** (1958), 31–36.
6. *Sur l'existence de solutions de certains nouveaux problèmes pour un système d'équations différentielles hyperboliques du second ordre à deux variables indépendantes*, Ann. Polon. Math. **4** (1957), 40–60.
7. *Sur l'existence d'une solution unique de certaines problèmes pour un système d'équations différentielles hyperboliques du second ordre à deux variables indépendantes*, Ann. Polon. Math. **4** (1958), 165–182.

A. Tychonoff

1. *Ein Fixpunktsatz*, Math. Ann. **111** (1935), 767–776.

B. van der Waerden

1. *Moderne Algebra*, 2nd Ed., Springer, Berlin, 1937.

J. von Neumann

1. *Über einen Hilfssatz der Variationsrechnung*, Abh. Math. Sem. Univ. Hamburg **8** (1931), 28–31.

INDEX

Affine mapping, 9
positive, negative, 9
Alexandroff and Hopf, 154, 186
Algebraic number of q-points, 27
Almost periodic solutions of quasilinear systems, 96
α-Hölder continuous, 157
Andronov and Chaikin, 75, 186
Andronov and Witt, 75, 186
a priori bound, 139
Ascoli's Theorem, 154
Autonomous quasilinear systems, 105

Banach, 123, 186
Banach Fixed Point Theorem, 141
Banach space, 122
Barrar, 162, 176, 178, 186
Bartle, 156, 186
Barycentric subdivision, 18
Basic Existence Theorem for ordinary differential equations, 57
Basis for a linear space, 121
for an infinite-dimensional Hilbert space, 122
Bernstein, S., 156, 166, 186
Bers, 142, 186
Bessel's inequality, 123
Birkhoff and Kellogg, 133, 186
Blackburn and Mann, 171, 186
Bogoliubov and Mitropolski, 96, 186
Botts, see McShane and Botts
Boundary of a chain, 6
Boundary point, viii
Bourgin, 55, 186
Branching equations (≡bifurcation equations ≡Verzweigungsgleichungen), 66
Brouwer Fixed Point Theorem, 2
Browder, 55, 120, 186
Buck, 33, 187

Cell, of order q, 4
oriented, 4
Cell complex, 16
Cesari, 56, 72, 187
Cesari method, 151, 180-185
Chaikin, see Andronov and Chaikin
Chain, 5
on a complex, 18
Chain mapping
induced by m subdivisions, 19
induced by a simplicial mapping, 19
Chains, sum of, 6
Characteristic exponent, 74
Ciliberts, 180, 187
Class Cm, 157
Closed set, viii
Closure of a set, viii
Coddington and Levinson, 62, 111, 187
Compact set, 130
Compact (≡completely continuous) transformation, 131
Complete metric space, 122
Completely continuous (≡compact) transformation, 131
Complex, cell, 16
simplicial, 17
oriented simplicial, 18
Component of an open set, viii
Computation of degree
in R^{2n}, 43
in the plane, 38
Conjugate space of a Banach space, 142
Conjugate transformation, 142
Connected set, viii
Continuation Theorem for ordinary differential equations, 61
Continuous function, ix
Contraction mapping, 141
Convex hull, 4
Convex set, 4, 57
Cordes, 170, 187
Courant and Hilbert, 152, 156, 178, 187
Critical point, 70
stable, 70
asymptotically stable, 71
Cronin, 115, 150, 156, 171, 187
Cycle, 7
bounding 7
Cycles, homologous, 8
Cylinder over a complex, 21

Day, 123, 187
Degenerate case (≡resonance case), 67
Degree (see local degree)
Diameter of a set, ix

Differentiability Theorem for ordinary differential equations, 63
Differential, 33
Distance between sets, ix
Domain, 57
Douglis and Nirenberg, 162, 188
Duff, 170, 188

Entrainment of frequency, 91
ϵ-approximation, 23
Euclidean n-space, vii
Example,
  in which local degree is zero, 94
  in which local degree is not defined, 95
  Kakutani's, 125
  Leray's, 128
Existence Theorem for a System with a Parameter, 62

Finite-dimensional linear manifold, 121
Finn and Gilbarg, 171, 188
Fixed point theorem,
  Brouwer, 2
  Schauder, 131
Friedman, 176, 178, 188
Fundamental matrix, 63

General position, 11, 12
Gevrey, 178, 188
Gilbarg, see Finn and Gilbarg
Gomory, 109, 114, 115, 188
Graves, see also Hildebrandt and Graves, 157, 162, 188
Guglielmo, 180, 188

Hadamard, 156, 188
Hale, 96, 151, 188
Halmos, 123, 188
Heinz, 170, 188
Hilbert, see Courant and Hilbert
Hilbert space, 122
Hildebrandt and Graves, 146, 188
Homeomorphism, ix
Homologous cycles, 8
Homotopy,
  definition, 25
  of compact transformations, 137
Hopf, E., 164, 188
Hopf, H., see Alexandroff and Hopf
Hurewicz, 62, 111, 189

Implicit Function Theorem for Banach Spaces, 145
Index
  of a fixed point, 52
  of a q-point, 27
  of a vector field, 53
Inner product, 122
Inner product space, 122
Integral equation
  of first kind, 152
  of second kind, 152
Intersection number
  of chains, 12
  of hyperplanes, 11
  of simplexes, 10
Interval,
  closed, vii
  open, vii
In-the-large Implicit Function Theorem, 51
Invariance under homotopy, 25, 31
Isolated q-point, 26

Jacobian, 33
Jordan and von Neumann, 123, 189
Jordan Canonical Form, 71

Kakutani, 55, 189
Kakutani's example, 125
Kellogg, see Birkhoff and Kellogg
Klee, 120, 124, 189
Krasno′selskiĭ, 156, 189
Krasno′selskiĭ and Rutickiĭ, 156, 189
Kronecker integral, 1, 50
Krzyzanski and Schauder, 180, 189
Kyner, 151, 189

LaSalle and Lefschetz, 70, 189
Lefschetz, see also LaSalle and Lefschetz, 54, 62, 70, 74, 114, 118, 189
Leimanis, see Minorsky and Leimanis
Leray, 140, 190
Leray's example, 128
Leray-Schauder degree, 119
  definition, 136
  properties, 136-139
  computation, 139
Leray Schauder-Nirenberg result, statement of, 156
Levinson, see also Coddington and Levinson, 117, 118, 189
Limit point of positive semi-orbit, 110
Limit point ($\equiv$cluster point$\equiv$accumulation point) of a set, viii
Linear functional 142

Linear manifold, 120
Linear space, 120
Lipschitz condition, 57
Local degree, 30, 31
Locally connected metric space, viii

MacDuffee, 68, 190
McShane and Botts, 154, 190
Malferrai, 179, 190
Malkin, 70, 72, 82, 96, 190
Malkin's Theorem (for almost periodic solutions), 97
Mann, see Blackburn and Mann
Marden, 99, 190
Maximum Principle, 164
Metric space, viii
  complete, 122
  locally connected, viii
  separable, viii
Mikolajska, 151, 190
Minorsky, 85, 190
Minorsky and Leimanis, 85, 190
Miranda, 162, 164, 190
Mitropolski, see Bogoliubov and Mitropolski

Nagumo, 1, 120, 190
Nash, 179, 190
Nirenberg, 156, 165, 167, 168, 191
Nitsche, 171, 191
Nonresonance case (≡nondegenerate case), 67
Norm, 121
Normed linear space, 121

Oleĭnik, 178, 191
1−1 function, ix
Open set, viii
Operator (≡transformation), linear, 124
Orbit
  definition, 109
  periodic, 110
Orbitally stable solution of a system of ordinary differential equations, 71
Order of p relative to φ and $\overline{K}$.
  rigorous definition, 24
  intuitive version of definition, 25
Order of p relative to z^{n-1}, 15
Orientation of a cell, 4
Oriented R^n, 10
Oriented simplicial complex, 18
Orthogonal, 122
Orthonormal sets, 122

Parallelogram equality, 123
Piskorek, 179, 191
Pogorzelski, 155, 179, 180, 191
Poincaré, 115, 191
Poincaré-Bendixson Theorem, 110
Poincaré-Bohl Theorem, 25, 32
Polyhedron, 17
Positive semi-orbit
  definition, 110
  limit point of, 110
Prodi, 179, 191
Product theorem for local degree, 36

q-point, isolated, 26
q-point, regular, 27
Quasilinear system, 64

Rado, 165, 192
Rado and Reichelderfer, 1, 192
Reduction Theorem, 51
Regular q-point, 27
Reichelderfer, see Rado and Reichelderfer
Resonance case (≡degenerate case), 67
Riesz, 142, 192
  index, 142
  transformation, 142
Riesz and Sz.-Nagy, 123, 142, 192
Rudin, 154, 192

Saddle function, 165
Sard, 33, 192
Schaefer, 119, 192
Schaefer's Theorem, 133
Schauder, see also Krzyzanski and Schauder, 119, 162, 165, 166, 180, 192
  Existence Theorem, 161
  Fixed Point Theorem, 131
Schmidt, 155, 192
Schmidt operator, 146, 155
Side of a cell, 4
Similar matrices, 71
Simplex, 9
Simplicial complex, 17
  oriented, 18
Simplicial decomposition, 17
Simplicial mapping, 19
  continuous, 20
Simplicial subdivision, 17
Sphere in a normed linear space, 122
Stability Theorem for a Critical Point
  First, 72
  Second, 74

Stability Theorem for a Periodic Solution
    First, 73
    Second, 74
Stable critical point, 70
    asymptotically stable, 71
Stable solution of a system of ordinary differential equations, 71
    asymptotically stable, 71
Stoker, 96, 192
Stone, 123, 192
Subdivision
    simplicial, 17
    barycentric, 18
Subharmonic oscillations, 85
Subspace, 121
Szmydt, 180, 192-3
Sz.-Nagy, see Riesz and Sz.-Nagy

Topological mapping, 2
Totally degenerate system, three-dimensional, 93
Transformation (≡operator), linear, 124
Triangle condition with constant Δ, 164
Tychonoff, 120, 193

van der Waerden, 45, 193
Variation of constants formula, 63
Vector field, 53
    index, 53
von Neumann, see also Jordan and von Neumann, 166, 193

Witt, see Andronov and Witt